Precalculus 1

An Investigation of Functions

Chapters 1-4

Edition 2.0

David Lippman
Melonie Rasmussen

This book is also available to read free online at
http://www.opentextbookstore.com/precalc/
If you want a printed copy, buying from the bookstore is cheaper than printing yourself.

In addition to these rights, we give explicit permission to remix small portions of this book (less than 10% cumulative) into works that are CC-BY, CC-BY-SA-NC, or GFDL licensed.

Selected exercises were remixed from *Precalculus* by D.H. Collingwood and K.D. Prince, originally licensed under the GNU Free Document License, with permission from the authors. These are marked in the book as [UW].

Portions of chapter 3 were remixed from *College Algebra* by Carl Stitz and Jeff Zeager, originally licensed under a Creative Commons Attribution ShareAlike Non-Commercial license, used with permission from the authors.

Portions of chapter 9 were remixed from work by Lara Michaels, and contains content remixed from *Precalculus* by OpenStax, originally licensed under a Creative Commons Attribution license. The original version is available online for free at OpenStax.org.

Cover photos by Ralph Morasch and David Lippman, of artwork by John Rogers.
Lituus, 2010. Dichromatic glass and aluminum.
Washington State Arts Commission in partnership with Pierce College

About the Authors

David Lippman received his master's degree in mathematics from Western Washington University and has been teaching at Pierce College since Fall 2000.

Melonie Rasmussen also received her master's degree in mathematics from Western Washington University and has been teaching at Pierce College since Fall 2002. Prior to this Melonie taught for the Puyallup School district for 6 years after receiving her teaching credentials from Pacific Lutheran University.

We have both been long time advocates of open learning, open materials, and basically any idea that will reduce the cost of education for students. It started by supporting the college's calculator rental program, and running a book loan scholarship program. Eventually the frustration with the escalating costs of commercial text books and the online homework systems that charged for access led them to take action.

First, David developed IMathAS, open source online math homework software that runs WAMAP.org and MyOpenMath.com. Through this platform, we became integral parts of a vibrant sharing and learning community of teachers from around Washington State that support and contribute to WAMAP. Our pioneering efforts, supported by dozens of other dedicated faculty and financial support from the WA-SBCTC, have led to a system used by thousands of students every quarter, saving hundreds of thousands of dollars over comparable commercial offerings.

David continued further and wrote his first open textbook, *Math in Society*, a math for liberal arts majors book, after being frustrated by students having to pay $100+ for a textbook for a terminal course. Together, frustrated by both cost and the style of commercial texts, we began writing *PreCalculus: An Investigation of Functions* in 2010. Since then, David has contributed to several other open texts.

Acknowledgements

We would like to thank the following for their generous support and feedback.

- The community of WAMAP users and developers for creating a majority of the homework content used in our online homework sets.

- Pierce College students in our Fall 2010 - Summer 2011 Math 141 and Math 142 classes for helping correct typos, identifying videos related to the homework, and being our willing test subjects.

- The Open Course Library Project for providing the support needed to produce a full course package for these courses.

- Mike Kenyon, Chris Willett, Tophe Anderson, and Vauhn Foster-Grahler for reviewing the course and giving feedback and suggestions.

- Our Pierce College colleagues for providing their suggestions.

- Tophe Anderson, James Gray, and Lawrence Morales for their feedback and suggestions in content and examples.

- Jeff Eldridge for extensive proofreading and suggestions for clarification.

- James Sousa for developing videos associated with the online homework.

- Kevin Dimond for his work on indexing the book and creating PowerPoint slides.

- Faculty at Green River Community College and the Maricopa College District for their feedback and suggestions.

- Lara Michaels for contributing the basis for a conics chapter.

- The dozens of instructors who have sent us typos or suggestions over the years.

Preface

Over the years, when reviewing books we found that many had been mainstreamed by the publishers in an effort to appeal to everyone, leaving them with very little character. There were only a handful of books that had the conceptual and application driven focus we liked, and most of those were lacking in other aspects we cared about, like providing students sufficient examples and practice of basic skills. The largest frustration, however, was the never ending escalation of cost and being forced into new editions every three years. We began researching open textbooks, however the ability for those books to be adapted, remixed, or printed were often limited by the types of licenses, or didn't approach the material the way we wanted.

This book is available online for free, in both Word and PDF format. You are free to change the wording, add materials and sections or take them away. We welcome feedback, comments and suggestions for future development at precalc@opentextbookstore.com. Additionally, if you add a section, chapter or problems, we would love to hear from you and possibly add your materials so everyone can benefit.

In writing this book, our focus was on the story of functions. We begin with function notation, a basic toolkit of functions, and the basic operation with functions: composition and transformation. Building from these basic functions, as each new family of functions is introduced we explore the important features of the function: its graph, domain and range, intercepts, and asymptotes. The exploration then moves to evaluating and solving equations involving the function, finding inverses, and culminates with modeling using the function.

The "rule of four" is integrated throughout - looking at the functions verbally, graphically, numerically, as well as algebraically. We feel that using the "rule of four" gives students the tools they need to approach new problems from various angles. Often the "story problems of life" do not always come packaged in a neat equation. Being able to think critically, see the parts and build a table or graph a trend, helps us change the words into meaningful and measurable functions that model the world around us.

There is nothing we hate more than a chapter on exponential equations that begins "Exponential functions are functions that have the form $f(x)=a^x$." As each family of functions is introduced, we motivate the topic by looking at how the function arises from life scenarios or from modeling. Also, we feel it is important that precalculus be the bridge in level of thinking between algebra and calculus. In algebra, it is common to see numerous examples with very similar homework exercises, encouraging the student to mimic the examples. Precalculus provides a link that takes students from the basic plug & chug of formulaic calculations towards building an understanding that equations and formulas have deeper meaning and purpose. While you will find examples and similar exercises for the basic skills in this book, you will also find examples of multistep problem solving along with exercises in multistep problem solving. Often times these exercises will not exactly mimic the exercises, forcing the students to employ their critical thinking skills and apply the skills they've learned to new situations. By developing students' critical thinking and problem solving skills this course prepares students for the rigors of Calculus.

While we followed a fairly standard ordering of material in the first half of the book, we took some liberties in the trig portion of the book. It is our opinion that there is no need to separate unit circle trig from triangle trig, and instead integrated them in the first chapter. Identities are introduced in the first chapter, and revisited throughout. Likewise, solving is introduced in the second chapter and revisited more extensively in the third chapter. As with the first part of the book, an emphasis is placed on motivating the concepts and on modeling and interpretation.

About the Second Edition

About 4 years and several minor typo revisions after the original release of this book, we started contemplating creating a second edition. We didn't want to change much; we've always found it very annoying when new editions change things just for the sake of making it seem different. However, in talking with instructors from around the country, we knew there were a few topics that we had left out that other schools need. We didn't want to suffer the same "content bloat" that many commercial books do, but we also wanted to make it easier for more schools to adopt open resources.

We put our plans for a new revision on hold after OpenStax started working on a precalculus book, using the first edition of this text as a base. After the final product came out, though, we felt it had strayed a bit far from our original vision. We had written this text, not to be an encyclopedic reference text, but to be a concise, easy-to-read, student-friendly approach to precalculus. We valued contextual motivation and conceptual understanding over procedural skills. Our book took, in places, a non-traditional approach to topics and content ordering. Ultimately, we decided to go ahead with this second edition.

The primary changes in the second edition are:
- New, higher resolution graphs throughout
- New sections added to Chapter 3:
    - 3.4 Factor theorem (includes long division of polynomials)
    - 3.5 Real zeros of polynomials (using rational roots theorem)
    - 3.6 Complex zeros of polynomials
- Coverage of oblique asymptotes added to the rational equations section (now 3.7)
- A new section 8.5 on dot product of vectors
- A new chapter 9 on conic sections

There were many additional refinements, some new examples added, and Try it Now answers expanded, but most of the book remains unchanged.

Instructor Resources

As part of the Washington Open Course Library project, we developed a full course package to accompany this text. The course shell was built for the IMathAS online homework platform, and is available for Washington State faculty at www.wamap.org and mirrored for others at www.myopenmath.com. It contains:

- Online homework for each section (algorithmically generated, free response), most with video help associated.
- Video lessons for each section. The videos were mostly created and selected by James Sousa, of Mathispower4u.
- A selection of printable class worksheets, activities, and handouts
- Support materials for an example course (does not include all sections):
    o Suggested syllabus and Day by day course guide
    o Instructor guide with lecture outlines and examples
    o Discussion forums
    o Diagnostic review
    o Chapter review problems
    o Sample quizzes and sample chapter exams

The course shell was designed to follow Quality Matters (QM) guidelines, but has not yet been formally reviewed.

Getting Started

To get started using this textbook and the online supplementary materials,
- Request an instructor account on WAMAP (in Washington) or MyOpenMath (outside Washington).
- Review the table of contents of the text, and compare it to your course outcomes or student learning objectives. Determine which sections you will need to cover, and which to omit. If there are topics in your outcomes that are not in the text, explore other sources like the Stitz/Zeager Precalc or OpenStax Precalc to supplement from. Also check the book's website, as we may offer additional online-only topics.
- Once your instructor account is approved, log in, and click **Add New Course**
- From the "Use content from a template course", select "Precalculus – Lippman/Rasmussen 2nd Ed". Note that you might also see two half-book templates, one covering chapters 1 – 4, and the other covering chapters 5 – 9.
- Once you have copied the course, go through and remove any sections you don't need for your course. Refer to the Training Course Quickstart videos in MyOpenMath and WAMAP for more details on how to make those changes.

How To Be Successful In This Course

This is not a high school math course, although for some of you the content may seem familiar. There are key differences to what you will learn here, how quickly you will be required to learn it and how much work will be required of you.

You will no longer be shown a technique and be asked to mimic it repetitively as the only way to prove learning. Not only will you be required to master the technique, but you will also be required to extend that knowledge to new situations and build bridges between the material at hand and the next topic, making the course highly cumulative.

As a rule of thumb, for each hour you spend in class, you should expect this course will require an average of 2 hours of out-of-class focused study. This means that some of you with a stronger background in mathematics may take less, but if you have a weaker background or any math anxiety it will take you more.

Notice how this is the equivalent of having a part time job, and if you are taking a fulltime load of courses as many college students do, this equates to more than a full time job. If you must work, raise a family and take a full load of courses all at the same time, we recommend that you get a head start & get organized as soon as possible. We also recommend that you spread out your learning into daily chunks and avoid trying to cram or learn material quickly before an exam.

To be prepared, read through the material before it is covered in class and note or highlight the material that is new or confusing. The instructor's lecture and activities should not be the first exposure to the material. As you read, test your understanding with the Try it Now problems in the book. If you can't figure one out, try again after class, and ask for help if you still can't get it.

As soon as possible after the class session recap the day's lecture or activities into a meaningful format to provide a third exposure to the material. You could summarize your notes into a list of key points, or reread your notes and try to work examples done in class without referring back to your notes. Next, begin any assigned homework. The next day, if the instructor provides the opportunity to clarify topics or ask questions, do not be afraid to ask. If you are afraid to ask, then you are not getting your money's worth! If the instructor does not provide this opportunity, be prepared to go to a tutoring center or build a peer study group. Put in quality effort and time and you can get quality results.

Lastly, if you feel like you do not understand a topic. Don't wait, ASK FOR HELP!

ASK: **A**sk a teacher or tutor, **S**earch for ancillaries, **K**eep a detailed list of questions
FOR: **F**ind additional resources, **O**rganize the material, **R**esearch other learning options
HELP: **H**ave a support network, **E**xamine your weaknesses, **L**ist specific examples & **P**ractice

Best of luck learning! We hope you like the course & love the price.
David & Melonie

Table of Contents

NOTE: This printed text only contains Chapters 1-4 of the book. The remainder of the book can be read online at http://www.opentextbookstore.com/precalc/ or purchased as a separated printed text.

Chapter 1: Functions

Section 1.1 Functions and Function Notation

What is a Function?

The natural world is full of relationships between quantities that change. When we see these relationships, it is natural for us to ask "If I know one quantity, can I then determine the other?" This establishes the idea of an input quantity, or independent variable, and a corresponding output quantity, or dependent variable. From this we get the notion of a functional relationship in which the output can be determined from the input.

For some quantities, like height and age, there are certainly relationships between these quantities. Given a specific person and any age, it is easy enough to determine their height, but if we tried to reverse that relationship and determine age from a given height, that would be problematic, since most people maintain the same height for many years.

Function
Function: A rule for a relationship between an input, or independent, quantity and an output, or dependent, quantity in which each input value uniquely determines one output value. We say "the output is a function of the input."

Example 1
In the height and age example above, is height a function of age? Is age a function of height? In the height and age example above, it would be correct to say that height is a function of age, since each age uniquely determines a height. For example, on my 18th birthday, I had exactly one height of 69 inches. However, age is not a function of height, since one height input might correspond with more than one output age. For example, for an input height of 70 inches, there is more than one output of age since I was 70 inches at the age of 20 and 21.

Example 2

At a coffee shop, the menu consists of items and their prices. Is price a function of the item? Is the item a function of the price?

We could say that price is a function of the item, since each input of an item has one output of a price corresponding to it. We could not say that item is a function of price, since two items might have the same price.

Example 3

In many classes the overall percentage you earn in the course corresponds to a decimal grade point. Is decimal grade a function of percentage? Is percentage a function of decimal grade?

For any percentage earned, there would be a decimal grade associated, so we could say that the decimal grade is a function of percentage. That is, if you input the percentage, your output would be a decimal grade. Percentage may or may not be a function of decimal grade, depending upon the teacher's grading scheme. With some grading systems, there are a range of percentages that correspond to the same decimal grade.

One-to-One Function

Sometimes in a relationship each input corresponds to exactly one output, and every output corresponds to exactly one input. We call this kind of relationship a **one-to-one function**.

From Example 3, *if* each unique percentage corresponds to one unique decimal grade point and each unique decimal grade point corresponds to one unique percentage then it is a one-to-one function.

Try it Now

Let's consider bank account information.

1. Is your balance a function of your bank account number?
   (if you input a bank account number does it make sense that the output is your balance?)

2. Is your bank account number a function of your balance?
   (if you input a balance does it make sense that the output is your bank account number?)

Function Notation

To simplify writing out expressions and equations involving functions, a simplified notation is often used. We also use descriptive variables to help us remember the meaning of the quantities in the problem.

Rather than write "height is a function of age", we could use the descriptive variable h to represent height and we could use the descriptive variable a to represent age.

"height is a function of age"   if we name the function f we write
"h is f of a"   or more simply
$h = f(a)$ we could instead name the function h and write
$h(a)$ which is read "h of a"

Remember we can use any variable to name the function; the notation $h(a)$ shows us that h depends on a. The value "a" must be put into the function "h" to get a result. Be careful - the parentheses indicate that age is input into the function (Note: do not confuse these parentheses with multiplication!).

Function Notation

The notation output $= f(\text{input})$ defines a function named f. This would be read "output is f of input"

Example 4

Introduce function notation to represent a function that takes as input the name of a month, and gives as output the number of days in that month.

The number of days in a month is a function of the name of the month, so if we name the function f, we could write "days $= f(\text{month})$" or $d = f(m)$. If we simply name the function d, we could write $d(m)$

For example, $d(\text{March}) = 31$, since March has 31 days. The notation $d(m)$ reminds us that the number of days, d (the output) is dependent on the name of the month, m (the input)

Example 5

A function $N = f(y)$ gives the number of police officers, N, in a town in year y. What does $f(2005) = 300$ tell us?

When we read $f(2005) = 300$, we see the input quantity is 2005, which is a value for the input quantity of the function, the year (y). The output value is 300, the number of police officers (N), a value for the output quantity. Remember $N=f(y)$. This tells us that in the year 2005 there were 300 police officers in the town.

Tables as Functions

Functions can be represented in many ways: Words (as we did in the last few examples), tables of values, graphs, or formulas. Represented as a table, we are presented with a list of input and output values.

In some cases, these values represent everything we know about the relationship, while in other cases the table is simply providing us a few select values from a more complete relationship.

Table 1: This table represents the input, number of the month (January = 1, February = 2, and so on) while the output is the number of days in that month. This represents everything we know about the months & days for a given year (that is not a leap year)

(input) Month number, m	1	2	3	4	5	6	7	8	9	10	11	12
(output) Days in month, D	31	28	31	30	31	30	31	31	30	31	30	31

Table 2: The table below defines a function $Q = g(n)$. Remember this notation tells us g is the name of the function that takes the input n and gives the output Q.

n	1	2	3	4	5
Q	8	6	7	6	8

Table 3: This table represents the age of children in years and their corresponding heights. This represents just some of the data available for height and ages of children.

(input) a, age in years	5	5	6	7	8	9	10
(output) h, height inches	40	42	44	47	50	52	54

Example 6

Which of these tables define a function (if any)? Are any of them one-to-one?

Input	Output
2	1
5	3
8	6

Input	Output
-3	5
0	1
4	5

Input	Output
1	0
5	2
5	4

The first and second tables define functions. In both, each input corresponds to exactly one output. The third table does not define a function since the input value of 5 corresponds with two different output values.

Only the first table is one-to-one; it is both a function, and each output corresponds to exactly one input. Although table 2 is a function, because each input corresponds to exactly one output, each output does not correspond to exactly one input so this function is not one-to-one. Table 3 is not even a function and so we don't even need to consider if it is a one-to-one function.

Try it Now

3. If each percentage earned translated to one letter grade, would this be a function? Is it one-to-one?

Solving and Evaluating Functions:

When we work with functions, there are two typical things we do: evaluate and solve. Evaluating a function is what we do when we know an input, and use the function to determine the corresponding output. Evaluating will always produce one result, since each input of a function corresponds to exactly one output.

Solving equations involving a function is what we do when we know an output, and use the function to determine the inputs that would produce that output. Solving a function could produce more than one solution, since different inputs can produce the same output.

Example 7

Using the table shown, where $Q=g(n)$

a) Evaluate $g(3)$

n	1	2	3	4	5
Q	8	6	7	6	8

Evaluating $g(3)$ (read: "g of 3")
means that we need to determine the output value, Q, of the function g given the input value of $n=3$. Looking at the table, we see the output corresponding to $n=3$ is $Q=7$, allowing us to conclude $g(3) = 7$.

b) Solve $g(n) = 6$

Solving $g(n) = 6$ means we need to determine what input values, n, produce an output value of 6. Looking at the table we see there are two solutions: $n = 2$ and $n = 4$.

When we input 2 into the function g, our output is $Q = 6$

When we input 4 into the function g, our output is also $Q = 6$

Try it Now
4. Using the function in Example 7, evaluate $g(4)$

Graphs as Functions

Oftentimes a graph of a relationship can be used to define a function. By convention, graphs are typically created with the input quantity along the horizontal axis and the output quantity along the vertical.

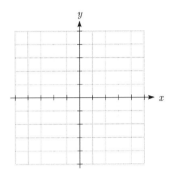

The most common graph has y on the vertical axis and x on the horizontal axis, and we say y is a function of x, or $y = f(x)$ when the function is named f.

Example 8

Which of these graphs defines a function $y=f(x)$? Which of these graphs defines a one-to-one function?

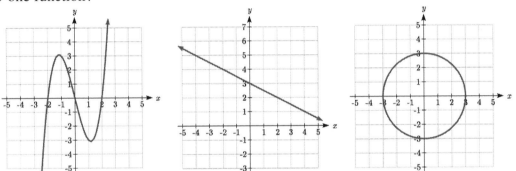

Looking at the three graphs above, the first two define a function $y=f(x)$, since for each input value along the horizontal axis there is exactly one output value corresponding, determined by the y-value of the graph. The 3rd graph does not define a function $y=f(x)$ since some input values, such as $x=2$, correspond with more than one output value.

Graph 1 is not a one-to-one function. For example, the output value 3 has two corresponding input values, -1 and 2.3

Graph 2 is a one-to-one function; each input corresponds to exactly one output, and every output corresponds to exactly one input.

Graph 3 is not even a function so there is no reason to even check to see if it is a one-to-one function.

Vertical Line Test

The **vertical line test** is a handy way to think about whether a graph defines the vertical output as a function of the horizontal input. Imagine drawing vertical lines through the graph. If any vertical line would cross the graph more than once, then the graph does not define only one vertical output for each horizontal input.

Horizontal Line Test

Once you have determined that a graph defines a function, an easy way to determine if it is a one-to-one function is to use the **horizontal line test**. Draw horizontal lines through the graph. If any horizontal line crosses the graph more than once, then the graph does not define a one-to-one function.

Evaluating a function using a graph requires taking the given input and using the graph to look up the corresponding output. Solving a function equation using a graph requires taking the given output and looking on the graph to determine the corresponding input.

Example 9

Given the graph of $f(x)$
a) Evaluate $f(2)$
b) Solve $f(x) = 4$

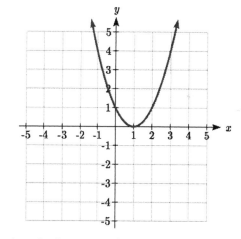

a) To evaluate $f(2)$, we find the input of $x=2$ on the horizontal axis. Moving up to the graph gives the point (2, 1), giving an output of $y=1$. $f(2) = 1$.

b) To solve $f(x) = 4$, we find the value 4 on the vertical axis because if $f(x) = 4$ then 4 is the output. Moving horizontally across the graph gives two points with the output of 4: (-1,4) and (3,4). These give the two solutions to $f(x) = 4$: $x = -1$ or $x = 3$
This means $f(-1)=4$ and $f(3)=4$, or when the input is -1 or 3, the output is 4.

Notice that while the graph in the previous example is a function, getting two input values for the output value of 4 shows us that this function is not one-to-one.

Try it Now
5. Using the graph from example 9, solve $f(x)=1$.

Formulas as Functions

When possible, it is very convenient to define relationships using formulas. If it is possible to express the output as a formula involving the input quantity, then we can define a function.

Example 10

Express the relationship $2n + 6p = 12$ as a function $p = f(n)$ if possible.

To express the relationship in this form, we need to be able to write the relationship where p is a function of n, which means writing it as $p =$ [something involving n].

$2n + 6p = 12$ subtract $2n$ from both sides
$6p = 12 - 2n$ divide both sides by 6 and simplify

$$p = \frac{12-2n}{6} = \frac{12}{6} - \frac{2n}{6} = 2 - \frac{1}{3}n$$

Having rewritten the formula as $p=$, we can now express p as a function:

$$p = f(n) = 2 - \frac{1}{3}n$$

It is important to note that not every relationship can be expressed as a function with a formula.

Note the important feature of an equation written as a function is that the output value can be determined directly from the input by doing evaluations - no further solving is required. This allows the relationship to act as a magic box that takes an input, processes it, and returns an output. Modern technology and computers rely on these functional relationships, since the evaluation of the function can be programmed into machines, whereas solving things is much more challenging.

Example 11

Express the relationship $x^2 + y^2 = 1$ as a function $y = f(x)$ if possible.

If we try to solve for y in this equation:
$$y^2 = 1 - x^2$$
$$y = \pm\sqrt{1-x^2}$$

We end up with two outputs corresponding to the same input, so this relationship cannot be represented as a single function $y = f(x)$.

As with tables and graphs, it is common to evaluate and solve functions involving formulas. Evaluating will require replacing the input variable in the formula with the value provided and calculating. Solving will require replacing the output variable in the formula with the value provided, and solving for the input(s) that would produce that output.

Example 12

Given the function $k(t) = t^3 + 2$
a) Evaluate $k(2)$
b) Solve $k(t) = 1$

a) To evaluate $k(2)$, we plug in the input value 2 into the formula wherever we see the input variable t, then simplify

$k(2) = 2^3 + 2$
$k(2) = 8 + 2$
So $k(2) = 10$

b) To solve $k(t) = 1$, we set the formula for $k(t)$ equal to 1, and solve for the input value that will produce that output

$k(t) = 1$ substitute the original formula $k(t) = t^3 + 2$
$t^3 + 2 = 1$ subtract 2 from each side
$t^3 = -1$ take the cube root of each side
$t = -1$

When solving an equation using formulas, you can check your answer by using your solution in the original equation to see if your calculated answer is correct.

We want to know if $k(t) = 1$ is true when $t = -1$.
$k(-1) = (-1)^3 + 2$
    $= -1 + 2$
    $= 1$ which was the desired result.

Example 13

Given the function $h(p) = p^2 + 2p$
a) Evaluate $h(4)$
b) Solve $h(p) = 3$

To evaluate $h(4)$ we substitute the value 4 for the input variable p in the given function.
a) $h(4) = (4)^2 + 2(4)$
    $= 16 + 8$
    $= 24$

b) $h(p) = 3$ Substitute the original function $h(p) = p^2 + 2p$

$p^2 + 2p = 3$ This is quadratic, so we can rearrange the equation to get it $= 0$

$p^2 + 2p - 3 = 0$ subtract 3 from each side

$p^2 + 2p - 3 = 0$ this is factorable, so we factor it

$(p + 3)(p - 1) = 0$

By the zero factor theorem since $(p + 3)(p - 1) = 0$, either $(p + 3) = 0$ or $(p - 1) = 0$ (or both of them equal 0) and so we solve both equations for p, finding $p = -3$ from the first equation and $p = 1$ from the second equation.

This gives us the solution: $h(p) = 3$ when $p = 1$ or $p = -3$

We found two solutions in this case, which tells us this function is not one-to-one.

Try it Now

6. Given the function $g(m) = \sqrt{m - 4}$
    a. Evaluate $g(5)$
    b. Solve $g(m) = 2$

Basic Toolkit Functions

In this text, we will be exploring functions – the shapes of their graphs, their unique features, their equations, and how to solve problems with them. When learning to read, we start with the alphabet. When learning to do arithmetic, we start with numbers. When working with functions, it is similarly helpful to have a base set of elements to build from. We call these our "toolkit of functions" – a set of basic named functions for which we know the graph, equation, and special features.

For these definitions we will use x as the input variable and $f(x)$ as the output variable.

Toolkit Functions
Linear

Constant: $f(x) = c$, where c is a constant (number)

Identity: $f(x) = x$

Absolute Value: $f(x) = |x|$

Power

Quadratic: $f(x) = x^2$

Cubic: $f(x) = x^3$

Reciprocal: $f(x) = \dfrac{1}{x}$

Reciprocal squared: $f(x) = \dfrac{1}{x^2}$

Square root: $f(x) = \sqrt[2]{x} = \sqrt{x}$

Cube root: $f(x) = \sqrt[3]{x}$

You will see these toolkit functions, combinations of toolkit functions, their graphs and their transformations frequently throughout this book. In order to successfully follow along later in the book, it will be very helpful if you can recognize these toolkit functions and their features quickly by name, equation, graph and basic table values.

Not every important equation can be written as $y = f(x)$. An example of this is the equation of a circle. Recall the distance formula for the distance between two points:

$$dist = \sqrt{(x_2 - x_1)^2 + (y_2 - y_1)^2}$$

A circle with radius r with center at (h, k) can be described as all points (x, y) a distance of r from the center, so using the distance formula, $r = \sqrt{(x-h)^2 + (y-k)^2}$, giving

Equation of a circle
A circle with radius r with center (h, k) has equation $r^2 = (x-h)^2 + (y-k)^2$

Graphs of the Toolkit Functions

Constant Function: $f(x) = 2$ Identity: $f(x) = x$

Absolute Value: $f(x) = |x|$

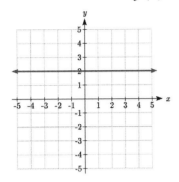
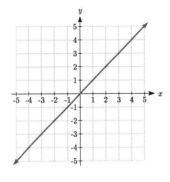
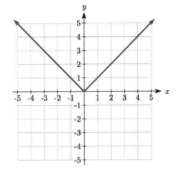

Quadratic: $f(x) = x^2$ Cubic: $f(x) = x^3$ Square root: $f(x) = \sqrt{x}$

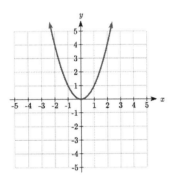
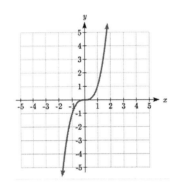
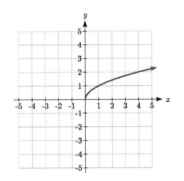

Cube root: $f(x) = \sqrt[3]{x}$ Reciprocal: $f(x) = \dfrac{1}{x}$ Reciprocal squared: $f(x) = \dfrac{1}{x^2}$

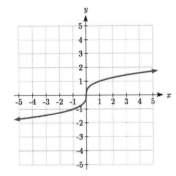
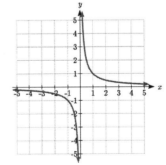
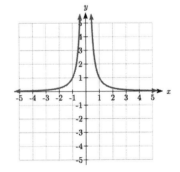

Important Topics of this Section
Definition of a function
Input (independent variable)
Output (dependent variable)
Definition of a one-to-one function
Function notation
Descriptive variables
Functions in words, tables, graphs & formulas
Vertical line test
Horizontal line test
Evaluating a function at a specific input value
Solving a function given a specific output value
Toolkit Functions

Try it Now Answers

1. Yes: for each bank account, there would be one balance associated

2. No: there could be several bank accounts with the same balance

3. Yes it's a function; No, it's not one-to-one (several percents give the same letter grade)

4. When $n=4$, $Q=g(4)=6$

5. There are two points where the output is 1: $x = 0$ or $x = 2$

6. a. $g(5) = \sqrt{5-4} = 1$

   b. $\sqrt{m-4} = 2$. Square both sides to get $m - 4 = 4$. $m = 8$

Section 1.1 Exercises

1. The amount of garbage, G, produced by a city with population p is given by
   $G = f(p)$. G is measured in tons per week, and p is measured in thousands of people.
   a. The town of Tola has a population of 40,000 and produces 13 tons of garbage
      each week. Express this information in terms of the function f.
   b. Explain the meaning of the statement $f(5) = 2$.

2. The number of cubic yards of dirt, D, needed to cover a garden with area a square
   feet is given by $D = g(a)$.
   a. A garden with area 5000 ft^2 requires 50 cubic yards of dirt. Express this
      information in terms of the function g.
   b. Explain the meaning of the statement $g(100) = 1$.

3. Let $f(t)$ be the number of ducks in a lake t years after 1990. Explain the meaning of
   each statement:
   a. $f(5) = 30$ \qquad b. $f(10) = 40$

4. Let $h(t)$ be the height above ground, in feet, of a rocket t seconds after launching.
   Explain the meaning of each statement:
   a. $h(1) = 200$ \qquad b. $h(2) = 350$

5. Select all of the following graphs which represent y as a function of x.

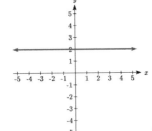

a

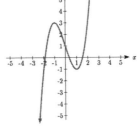

b

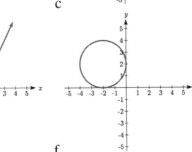

c

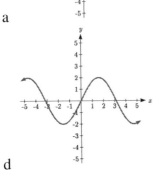

d

e

f

6. Select all of the following graphs which represent *y* as a function of *x*.

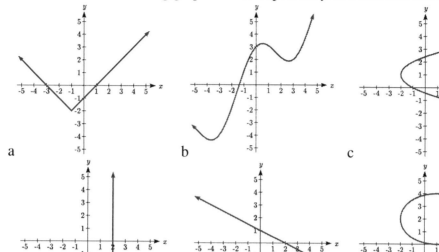

a   b   c

d   e   f

7. Select all of the following tables which represent *y* as a function of *x*.

a.
x	5	10	15
y	3	8	14

b.
x	5	10	15
y	3	8	8

c.
x	5	10	10
y	3	8	14

8. Select all of the following tables which represent *y* as a function of *x*.

a.
x	2	6	13
y	3	10	10

b.
x	2	6	6
y	3	10	14

c.
x	2	6	13
y	3	10	14

9. Select all of the following tables which represent *y* as a function of *x*.

a.
x	y
0	-2
3	1
4	6
8	9
3	1

b.
x	y
-1	-4
2	3
5	4
8	7
12	11

c.
x	y
0	-5
3	1
3	4
9	8
16	13

d.
x	y
-1	-4
1	2
4	2
9	7
12	13

10. Select all of the following tables which represent *y* as a function of *x*.

a.
x	y
-4	-2
3	2
6	4
9	7
12	16

b.
x	y
-5	-3
2	1
2	4
7	9
11	10

c.
x	y
-1	-3
1	2
5	4
9	8
1	2

d.
x	y
-1	-5
3	1
5	1
8	7
14	12

11. Select all of the following tables which represent y as a function of x **and** are one-to-one.

a.
x	3	8	12
y	4	7	7

b.
x	3	8	12
y	4	7	13

c.
x	3	8	8
y	4	7	13

12. Select all of the following tables which represent y as a function of x **and** are one-to-one.

a.
x	2	8	8
y	5	6	13

b.
x	2	8	14
y	5	6	6

c.
x	2	8	14
y	5	6	13

13. Select all of the following graphs which are **one-to-one functions**.

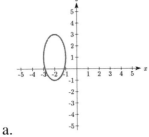

a.

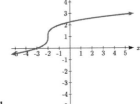

b.

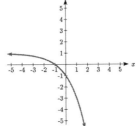

c.

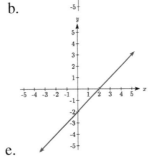

d.

e.

f.

14. Select all of the following graphs which are **one-to-one functions**.

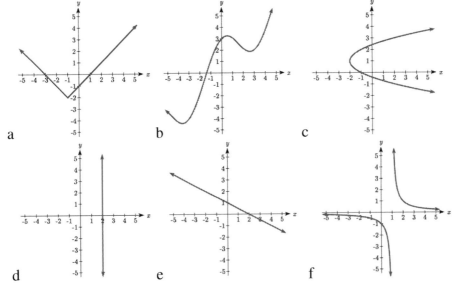

a

b

c

d

e

f

Given each function $f(x)$ graphed, evaluate $f(1)$ and $f(3)$

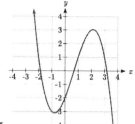

15.

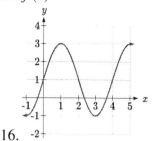

16.

17. Given the function $g(x)$ graphed here,
   a.   Evaluate $g(2)$
   b.   Solve $g(x)=2$

18. Given the function $f(x)$ graphed here.
   a.   Evaluate $f(4)$
   b.   Solve $f(x)=4$

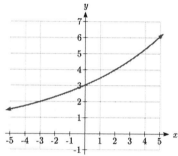

19. Based on the table below,
   a. Evaluate $f(3)$ b. Solve $f(x)=1$

x	0	1	2	3	4	5	6	7	8	9
$f(x)$	74	28	1	53	56	3	36	45	14	47

20. Based on the table below,
   a. Evaluate $f(8)$ b. Solve $f(x)=7$

x	0	1	2	3	4	5	6	7	8	9
$f(x)$	62	8	7	38	86	73	70	39	75	34

For each of the following functions, evaluate: $f(-2)$, $f(-1)$, $f(0)$, $f(1)$, and $f(2)$

21. $f(x)=4-2x$

22. $f(x)=8-3x$

23. $f(x)=8x^2-7x+3$

24. $f(x)=6x^2-7x+4$

25. $f(x)=-x^3+2x$

26. $f(x)=5x^4+x^2$

27. $f(x)=3+\sqrt{x+3}$

28. $f(x)=4-\sqrt[3]{x-2}$

29. $f(x)=(x-2)(x+3)$

30. $f(x)=(x+3)(x-1)^2$

31. $f(x)=\dfrac{x-3}{x+1}$

32. $f(x)=\dfrac{x-2}{x+2}$

33. $f(x)=2^x$

34. $f(x)=3^x$

35. Suppose $f(x) = x^2 + 8x - 4$. Compute the following:

      a. $f(-1) + f(1)$ b. $f(-1) - f(1)$

36. Suppose $f(x) = x^2 + x + 3$. Compute the following:

      a. $f(-2) + f(4)$ b. $f(-2) - f(4)$

37. Let $f(t) = 3t + 5$

      a. Evaluate $f(0)$ b. Solve $f(t) = 0$

38. Let $g(p) = 6 - 2p$

      a. Evaluate $g(0)$ b. Solve $g(p) = 0$

39. Match each function name with its equation.

  a. $y = x$ i.   Cube root

  b. $y = x^3$ ii.   Reciprocal

  c. $y = \sqrt[3]{x}$ iii.   Linear

                         iv.   Square Root

  d. $y = \dfrac{1}{x}$ v.   Absolute Value

                         vi.   Quadratic

  e. $y = x^2$ vii.   Reciprocal Squared

  f. $y = \sqrt{x}$ viii. Cubic

  g. $y = |x|$

  h. $y = \dfrac{1}{x^2}$

40. Match each graph with its equation.

  a. $y = x$ i.   ii.   iii.   iv.

  b. $y = x^3$

  c. $y = \sqrt[3]{x}$

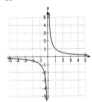 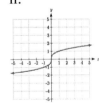 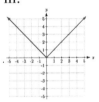 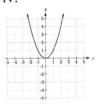

  d. $y = \dfrac{1}{x}$

  e. $y = x^2$

  f. $y = \sqrt{x}$ v.   vi.   vii.   viii.

  g. $y = |x|$

  h. $y = \dfrac{1}{x^2}$

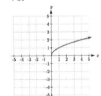 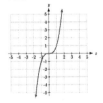 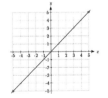

41. Match each table with its equation.

    a. $y = x^2$
    b. $y = x$
    c. $y = \sqrt{x}$
    d. $y = 1/x$
    e. $y = |x|$
    f. $y = x^3$

i.

In	Out
-2	-0.5
-1	-1
0	_
1	1
2	0.5
3	0.33

ii.

In	Out
-2	-2
-1	-1
0	0
1	1
2	2
3	3

iii.

In	Out
-2	-8
-1	-1
0	0
1	1
2	8
3	27

iv.

In	Out
-2	4
-1	1
0	0
1	1
2	4
3	9

v.

In	Out
-2	_
-1	_
0	0
1	1
4	2
9	3

vi.

In	Out
-2	2
-1	1
0	0
1	1
2	2
3	3

42. Match each equation with its table

    a. Quadratic
    b. Absolute Value
    c. Square Root
    d. Linear
    e. Cubic
    f. Reciprocal

i.

In	Out
-2	-0.5
-1	-1
0	_
1	1
2	0.5
3	0.33

ii.

In	Out
-2	-2
-1	-1
0	0
1	1
2	2
3	3

iii.

In	Out
-2	-8
-1	-1
0	0
1	1
2	8
3	27

iv.

In	Out
-2	4
-1	1
0	0
1	1
2	4
3	9

v.

In	Out
-2	_
-1	_
0	0
1	1
4	2
9	3

vi.

In	Out
-2	2
-1	1
0	0
1	1
2	2
3	3

43. Write the equation of the circle centered at $(3, -9)$ with radius 6.

44. Write the equation of the circle centered at $(9, -8)$ with radius 11.

45. Sketch a reasonable graph for each of the following functions. [UW]
    a. Height of a person depending on age.
    b. Height of the top of your head as you jump on a pogo stick for 5 seconds.
    c. The amount of postage you must put on a first class letter, depending on the weight of the letter.

46. Sketch a reasonable graph for each of the following functions. [UW]
    a. Distance of your big toe from the ground as you ride your bike for 10 seconds.
    b. Your height above the water level in a swimming pool after you dive off the high board.
    c. The percentage of dates and names you'll remember for a history test, depending on the time you study.

47. Using the graph shown,
    a. Evaluate $f(c)$
    b. Solve $f(x) = p$
    c. Suppose $f(b) = z$. Find $f(z)$
    d. What are the coordinates of points L and K?

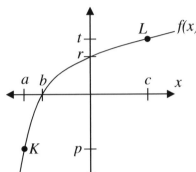

48. Dave leaves his office in Padelford Hall on his way to teach in Gould Hall. Below are several different scenarios. In each case, sketch a plausible (reasonable) graph of the function $s = d(t)$ which keeps track of Dave's distance s from Padelford Hall at time t. Take distance units to be "feet" and time units to be "minutes." Assume Dave's path to Gould Hall is long a straight line which is 2400 feet long. [UW]

    a. Dave leaves Padelford Hall and walks at a constant spend until he reaches Gould Hall 10 minutes later.

    b. Dave leaves Padelford Hall and walks at a constant speed. It takes him 6 minutes to reach the half-way point. Then he gets confused and stops for 1 minute. He then continues on to Gould Hall at the same constant speed he had when he originally left Padelford Hall.

    c. Dave leaves Padelford Hall and walks at a constant speed. It takes him 6 minutes to reach the half-way point. Then he gets confused and stops for 1 minute to figure out where he is. Dave then continues on to Gould Hall at twice the constant speed he had when he originally left Padelford Hall.

d. Dave leaves Padelford Hall and walks at a constant speed. It takes him 6 minutes to reach the half-way point. Then he gets confused and stops for 1 minute to figure out where he is. Dave is totally lost, so he simply heads back to his office, walking the same constant speed he had when he originally left Padelford Hall.

e. Dave leaves Padelford heading for Gould Hall at the same instant Angela leaves Gould Hall heading for Padelford Hall. Both walk at a constant speed, but Angela walks twice as fast as Dave. Indicate a plot of "distance from Padelford" vs. "time" for the both Angela and Dave.

f. Suppose you want to sketch the graph of a new function s = g(t) that keeps track of Dave's distance s from Gould Hall at time t. How would your graphs change in (a)-(e)?

Section 1.2 Domain and Range

One of our main goals in mathematics is to model the real world with mathematical functions. In doing so, it is important to keep in mind the limitations of those models we create.

This table shows a relationship between circumference and height of a tree as it grows.

Circumference, c	1.7	2.5	5.5	8.2	13.7
Height, h	24.5	31	45.2	54.6	92.1

While there is a strong relationship between the two, it would certainly be ridiculous to talk about a tree with a circumference of -3 feet, or a height of 3000 feet. When we identify limitations on the inputs and outputs of a function, we are determining the domain and range of the function.

Domain and Range

Domain: The set of possible input values to a function

Range: The set of possible output values of a function

Example 1

Using the tree table above, determine a reasonable domain and range.

We could combine the data provided with our own experiences and reason to approximate the domain and range of the function $h = f(c)$. For the domain, possible values for the input circumference c, it doesn't make sense to have negative values, so $c > 0$. We could make an educated guess at a maximum reasonable value, or look up that the maximum circumference measured is about 119 feet[1]. With this information, we would say a reasonable domain is $0 < c \leq 119$ feet.

Similarly for the range, it doesn't make sense to have negative heights, and the maximum height of a tree could be looked up to be 379 feet, so a reasonable range is $0 < h \leq 379$ feet.

---

[1] http://en.wikipedia.org/wiki/Tree, retrieved July 19, 2010

Example 2

When sending a letter through the United States Postal Service, the price depends upon the weight of the letter[2], as shown in the table below. Determine the domain and range.

Letters	
Weight not Over	Price
1 ounce	$0.44
2 ounces	$0.61
3 ounces	$0.78
3.5 ounces	$0.95

Suppose we notate Weight by w and Price by p, and set up a function named P, where Price, p is a function of Weight, w. $p = P(w)$.

Since acceptable weights are 3.5 ounces or less, and negative weights don't make sense, the domain would be $0 < w \leq 3.5$. Technically 0 could be included in the domain, but logically it would mean we are mailing nothing, so it doesn't hurt to leave it out.

Since possible prices are from a limited set of values, we can only define the range of this function by listing the possible values. The range is $p = \$0.44, \$0.61, \$0.78$, or $\$0.95$.

Try it Now

1. The population of a small town in the year 1960 was 100 people. Since then the population has grown to 1400 people reported during the 2010 census. Choose descriptive variables for your input and output and use interval notation to write the domain and range.

Notation

In the previous examples, we used inequalities to describe the domain and range of the functions. This is one way to describe intervals of input and output values, but is not the only way. Let us take a moment to discuss notation for domain and range.

Using inequalities, such as $0 < c \leq 163$, $0 < w \leq 3.5$, and $0 < h \leq 379$ imply that we are interested in all values between the low and high values, including the high values in these examples.

However, occasionally we are interested in a specific list of numbers like the range for the price to send letters, $p = \$0.44, \$0.61, \$0.78$, or $\$0.95$. These numbers represent a set of specific values: $\{0.44, 0.61, 0.78, 0.95\}$.

---

[2] http://www.usps.com/prices/first-class-mail-prices.htm, retrieved July 19, 2010

Representing values as a set, or giving instructions on how a set is built, leads us to another type of notation to describe the domain and range.

Suppose we want to describe the values for a variable x that are 10 or greater, but less than 30. In inequalities, we would write $10 \le x < 30$.

When describing domains and ranges, we sometimes extend this into **set-builder notation**, which would look like this: $\{x \mid 10 \le x < 30\}$. The curly brackets { } are read as "the set of", and the vertical bar | is read as "such that", so altogether we would read $\{x \mid 10 \le x < 30\}$ as "the set of x-values such that 10 is less than or equal to x and x is less than 30."

When describing ranges in set-builder notation, we could similarly write something like $\{f(x) \mid 0 < f(x) < 100\}$, or if the output had its own variable, we could use it. So for our tree height example above, we could write for the range $\{h \mid 0 < h \le 379\}$. In set-builder notation, if a domain or range is not limited, we could write $\{t \mid t \text{ is a real number}\}$, or $\{t \mid t \in \mathbb{R}\}$, read as "the set of t-values such that t is an element of the set of real numbers.

A more compact alternative to set-builder notation is **interval notation**, in which intervals of values are referred to by the starting and ending values. Curved parentheses are used for "strictly less than," and square brackets are used for "less than or equal to." Since infinity is not a number, we can't include it in the interval, so we always use curved parentheses with ∞ and $-\infty$. The table below will help you see how inequalities correspond to set-builder notation and interval notation:

Inequality	Set Builder Notation	Interval notation
$5 < h \le 10$	$\{h \mid 5 < h \le 10\}$	(5, 10]
$5 \le h < 10$	$\{h \mid 5 \le h < 10\}$	[5, 10)
$5 < h < 10$	$\{h \mid 5 < h < 10\}$	(5, 10)
$h < 10$	$\{h \mid h < 10\}$	$(-\infty, 10)$
$h \ge 10$	$\{h \mid h \ge 10\}$	$[10, \infty)$
all real numbers	$\{h \mid h \in \mathbb{R}\}$	$(-\infty, \infty)$

To combine two intervals together, using inequalities or set-builder notation we can use the word "or". In interval notation, we use the union symbol, $\cup$, to combine two unconnected intervals together.

Example 3

Describe the intervals of values shown on the line graph below using set builder and interval notations.

To describe the values, x, that lie in the intervals shown above we would say, "x is a real number greater than or equal to 1 and less than or equal to 3, or a real number greater than 5."

As an inequality it is: $1 \le x \le 3$ or $x > 5$

In set builder notation: $\{x \mid 1 \le x \le 3 \text{ or } x > 5\}$

In interval notation: $[1,3] \cup (5,\infty)$

Remember when writing or reading interval notation:
Using a square bracket [means the start value is included in the set
Using a parenthesis (means the start value is not included in the set

Try it Now

2. Given the following interval, write its meaning in words, set builder notation, and interval notation.

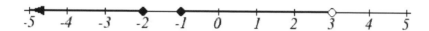

Domain and Range from Graphs

We can also talk about domain and range based on graphs. Since domain refers to the set of possible input values, the domain of a graph consists of all the input values shown on the graph. Remember that input values are almost always shown along the horizontal axis of the graph. Likewise, since range is the set of possible output values, the range of a graph we can see from the possible values along the vertical axis of the graph.

Be careful – if the graph continues beyond the window on which we can see the graph, the domain and range might be larger than the values we can see.

Example 4

Determine the domain and range of the graph below.

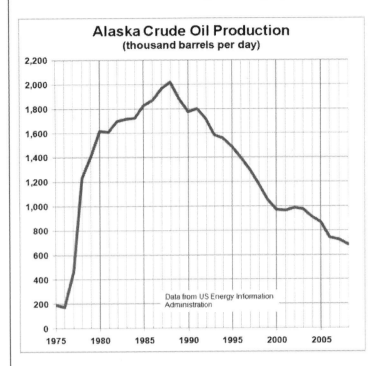

In the graph above[3], the input quantity along the horizontal axis appears to be "year", which we could notate with the variable *y*. The output is "thousands of barrels of oil per day", which we might notate with the variable *b*, for barrels. The graph would likely continue to the left and right beyond what is shown, but based on the portion of the graph that is shown to us, we can determine the domain is $1975 \le y \le 2008$, and the range is approximately $180 \le b \le 2010$.

In interval notation, the domain would be [1975, 2008] and the range would be about [180, 2010]. For the range, we have to approximate the smallest and largest outputs since they don't fall exactly on the grid lines.

Remember that, as in the previous example, *x* and *y* are not always the input and output variables. Using descriptive variables is an important tool to remembering the context of the problem.

Try it Now

3. Given the graph below write the domain and range in interval notation

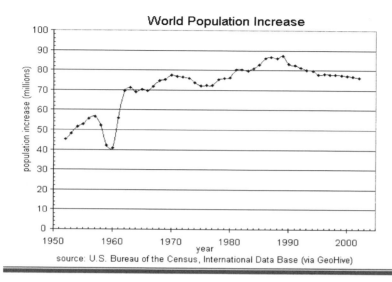

Domains and Ranges of the Toolkit functions

We will now return to our set of toolkit functions to note the domain and range of each.

Constant Function: $f(x) = c$
The domain here is not restricted; x can be anything. When this is the case we say the domain is all real numbers. The outputs are limited to the constant value of the function.
Domain: $(-\infty, \infty)$

Range: $[c]$
Since there is only one output value, we list it by itself in square brackets.

Identity Function: $f(x) = x$
Domain: $(-\infty, \infty)$
Range: $(-\infty, \infty)$

Quadratic Function: $f(x) = x^2$
Domain: $(-\infty, \infty)$
Range: $[0, \infty)$
Multiplying a negative or positive number by itself can only yield a positive output.

Cubic Function: $f(x) = x^3$
Domain: $(-\infty, \infty)$
Range: $(-\infty, \infty)$

Reciprocal: $f(x) = \dfrac{1}{x}$

Domain: $(-\infty, 0) \cup (0, \infty)$

Range: $(-\infty, 0) \cup (0, \infty)$

We cannot divide by 0 so we must exclude 0 from the domain.
One divide by any value can never be 0, so the range will not include 0.

Reciprocal squared: $f(x) = \dfrac{1}{x^2}$

Domain: $(-\infty, 0) \cup (0, \infty)$

Range: $(0, \infty)$

We cannot divide by 0 so we must exclude 0 from the domain.

Cube Root: $f(x) = \sqrt[3]{x}$

Domain: $(-\infty, \infty)$

Range: $(-\infty, \infty)$

Square Root: $f(x) = \sqrt[2]{x}$, commonly just written as, $f(x) = \sqrt{x}$

Domain: $[0, \infty)$

Range: $[0, \infty)$

When dealing with the set of real numbers we cannot take the square root of a negative number so the domain is limited to 0 or greater.

Absolute Value Function: $f(x) = |x|$

Domain: $(-\infty, \infty)$

Range: $[0, \infty)$

Since absolute value is defined as a distance from 0, the output can only be greater than or equal to 0.

Example 5

Find the domain of each function: a) $f(x) = 2\sqrt{x+4}$ b) $g(x) = \dfrac{3}{6-3x}$

a) Since we cannot take the square root of a negative number, we need the inside of the square root to be non-negative.
$x + 4 \geq 0$ when $x \geq -4$.
The domain of *f(x)* is $[-4, \infty)$.

b) We cannot divide by zero, so we need the denominator to be non-zero.
$6 - 3x = 0$ when $x = 2$, so we must exclude 2 from the domain.
The domain of *g(x)* is $(-\infty, 2) \cup (2, \infty)$.

Piecewise Functions

In the toolkit functions we introduced the absolute value function $f(x) = |x|$.

With a domain of all real numbers and a range of values greater than or equal to 0, the absolute value can be defined as the magnitude or modulus of a number, a real number value regardless of sign, the size of the number, or the distance from 0 on the number line. All of these definitions require the output to be greater than or equal to 0.

If we input 0, or a positive value the output is unchanged
$$f(x) = x \quad \text{if} \quad x \geq 0$$

If we input a negative value the sign must change from negative to positive.
$$f(x) = -x \quad \text{if} \quad x < 0, \quad \text{since multiplying a negative value by -1 makes it positive.}$$

Since this requires two different processes or pieces, the absolute value function is often called the most basic piecewise defined function.

Piecewise Function

A **piecewise function** is a function in which the formula used depends upon the domain the input lies in. We notate this idea like:

$$f(x) = \begin{cases} \text{formula 1} & \text{if} & \text{domain to use formula 1} \\ \text{formula 2} & \text{if} & \text{domain to use formula 2} \\ \text{formula 3} & \text{if} & \text{domain to use formula 3} \end{cases}$$

Example 6

A museum charges \$5 per person for a guided tour with a group of 1 to 9 people, or a fixed \$50 fee for 10 or more people in the group. Set up a function relating the number of people, n, to the cost, C.

To set up this function, two different formulas would be needed. $C = 5n$ would work for n values under 10, and $C = 50$ would work for values of n ten or greater. Notating this:
$$C(n) = \begin{cases} 5n & \text{if} & 0 < n < 10 \\ 50 & \text{if} & n \geq 10 \end{cases}$$

Example 7

A cell phone company uses the function below to determine the cost, C, in dollars for g gigabytes of data transfer.

$$C(g) = \begin{cases} 25 & \text{if} & 0 < g < 2 \\ 25 + 10(g-2) & \text{if} & g \geq 2 \end{cases}$$

Find the cost of using 1.5 gigabytes of data, and the cost of using 4 gigabytes of data.

To find the cost of using 1.5 gigabytes of data, $C(1.5)$, we first look to see which piece of domain our input falls in. Since 1.5 is less than 2, we use the first formula, giving $C(1.5) = \$25$.

The find the cost of using 4 gigabytes of data, $C(4)$, we see that our input of 4 is greater than 2, so we'll use the second formula. $C(4) = 25 + 10(4-2) = \$45$.

Example 8

Sketch a graph of the function $f(x) = \begin{cases} x^2 & \text{if} & x \leq 1 \\ 3 & \text{if} & 1 < x \leq 2 \\ 6 - x & \text{if} & x > 2 \end{cases}$

The first two component functions are from our library of Toolkit functions, so we know their shapes. We can imagine graphing each function, then limiting the graph to the indicated domain. At the endpoints of the domain, we put open circles to indicate where the endpoint is not included, due to a strictly-less-than inequality, and a closed circle where the endpoint is included, due to a less-than-or-equal-to inequality.

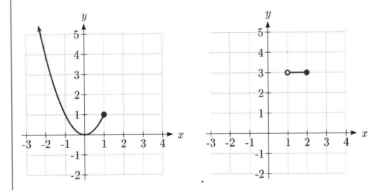

For the third function, you should recognize this as a linear equation from your previous coursework. If you remember how to graph a line using slope and intercept, you can do that. Otherwise, we could calculate a couple values, plot points, and connect them with a line.

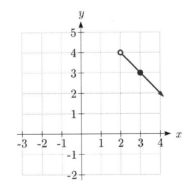

At $x = 2, f(2) = 6 - 2 = 4$. We place an open circle here.
At $x = 3, f(3) = 6 - 3 = 3$. Connect these points with a line.

Now that we have each piece individually, we combine them onto the same graph:

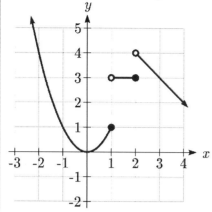

Try it Now

4. At Pierce College during the 2009-2010 school year tuition rates for in-state residents were $89.50 per credit for the first 10 credits, $33 per credit for credits 11-18, and for over 18 credits the rate is $73 per credit[4]. Write a piecewise defined function for the total tuition, T, at Pierce College during 2009-2010 as a function of the number of credits taken, c. Be sure to consider a reasonable domain and range.

Important Topics of this Section
Definition of domain
Definition of range
Inequalities
Interval notation
Set builder notation
Domain and Range from graphs
Domain and Range of toolkit functions
Piecewise defined functions

---

[4] https://www.pierce.ctc.edu/dist/tuition/ref/files/0910_tuition_rate.pdf, retrieved August 6, 2010

Try it Now Answers

1. Domain; y = years [1960,2010] ; Range, p = population, [100,1400]

2. a. Values that are less than or equal to -2, or values that are greater than or equal to -1 and less than 3

b. $\{x \,|\, x \le -2 \; or -1 \le x < 3\}$

c. $(-\infty, -2] \cup [-1, 3)$

3. Domain; y=years, [1952,2002] ; Range, p=population in millions, [40,88]

4. $T(c) = \begin{cases} 89.5c & if & c \le 10 \\ 895 + 33(c-10) & if & 10 < c \le 18 \\ 1159 + 73(c-18) & if & c > 18 \end{cases}$ Tuition, T, as a function of credits, c.

Reasonable domain should be whole numbers 0 to (answers may vary), e.g. $[0, 23]$

Reasonable range should be \$0 – (answers may vary), e.g. $[0,1524]$

Section 1.2 Exercises

Write the domain and range of the function using interval notation.

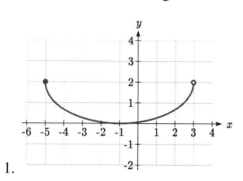

1.

2.

Write the domain and range of each graph as an inequality.

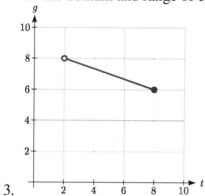

3.

4.

Suppose that you are holding your toy submarine under the water. You release it and it begins to ascend. The graph models the depth of the submarine as a function of time, stopping once the sub surfaces. What is the domain and range of the function in the graph?

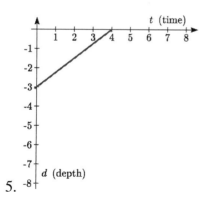

5.

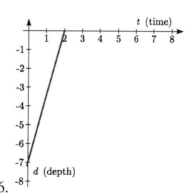

6.

Find the domain of each function

7. $f(x)=3\sqrt{x-2}$

8. $f(x)=5\sqrt{x+3}$

9. $f(x)=3-\sqrt{6-2x}$

10. $f(x)=5-\sqrt{10-2x}$

11. $f(x)=\dfrac{9}{x-6}$

12. $f(x)=\dfrac{6}{x-8}$

13. $f(x)=\dfrac{3x+1}{4x+2}$

14. $f(x)=\dfrac{5x+3}{4x-1}$

15. $f(x)=\dfrac{\sqrt{x+4}}{x-4}$

16. $f(x)=\dfrac{\sqrt{x+5}}{x-6}$

17. $f(x)=\dfrac{x-3}{x^2+9x-22}$

18. $f(x)=\dfrac{x-8}{x^2+8x-9}$

Given each function, evaluate: $f(-1)$, $f(0)$, $f(2)$, $f(4)$

19. $f(x)=\begin{cases}7x+3 & if \quad x<0\\7x+6 & if \quad x\geq0\end{cases}$

20. $f(x)=\begin{cases}4x-9 & if \quad x<0\\4x-18 & if \quad x\geq0\end{cases}$

21. $f(x)=\begin{cases}x^2-2 & if \quad x<2\\4+|x-5| & if \quad x\geq2\end{cases}$

22. $f(x)=\begin{cases}4-x^3 & if \quad x<1\\\sqrt{x+1} & if \quad x\geq1\end{cases}$

23. $f(x)=\begin{cases}5x & if \quad x<0\\3 & if \quad 0\leq x\leq3\\x^2 & if \quad x>3\end{cases}$

24. $f(x)=\begin{cases}x^3+1 & if \quad x<0\\4 & if \quad 0\leq x\leq3\\3x+1 & if \quad x>3\end{cases}$

Write a formula for the piecewise function graphed below.

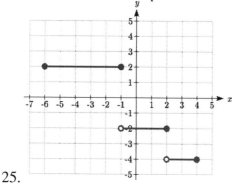

25.

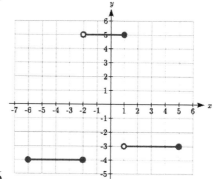

26.

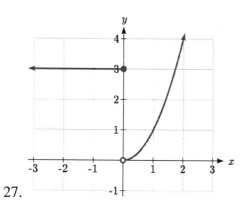

27.

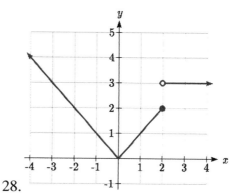

28.

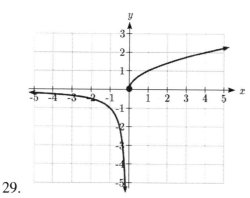

29.

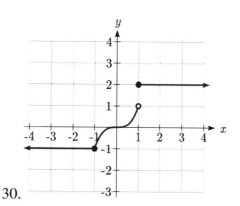

30.

Sketch a graph of each piecewise function

31. $f(x) = \begin{cases} |x| & if \quad x < 2 \\ 5 & if \quad x \geq 2 \end{cases}$

32. $f(x) = \begin{cases} 4 & if \quad x < 0 \\ \sqrt{x} & if \quad x \geq 0 \end{cases}$

33. $f(x) = \begin{cases} x^2 & if \quad x < 0 \\ x + 2 & if \quad x \geq 0 \end{cases}$

34. $f(x) = \begin{cases} x + 1 & if \quad x < 1 \\ x^3 & if \quad x \geq 1 \end{cases}$

35. $f(x) = \begin{cases} 3 & if \quad x \leq -2 \\ -x + 1 & if \quad -2 < x \leq 1 \\ 3 & if \quad x > 1 \end{cases}$

36. $f(x) = \begin{cases} -3 & if \quad x \leq -2 \\ x - 1 & if \quad -2 < x \leq 2 \\ 0 & if \quad x > 2 \end{cases}$

Section 1.3 Rates of Change and Behavior of Graphs

Since functions represent how an output quantity varies with an input quantity, it is natural to ask about the rate at which the values of the function are changing.

For example, the function $C(t)$ below gives the average cost, in dollars, of a gallon of gasoline t years after 2000.

t	2	3	4	5	6	7	8	9
$C(t)$	1.47	1.69	1.94	2.30	2.51	2.64	3.01	2.14

If we were interested in how the gas prices had changed between 2002 and 2009, we could compute that the cost per gallon had increased from $1.47 to $2.14, an increase of $0.67. While this is interesting, it might be more useful to look at how much the price changed *per year*. You are probably noticing that the price didn't change the same amount each year, so we would be finding the **average rate of change** over a specified amount of time.

The gas price increased by $0.67 from 2002 to 2009, over 7 years, for an average of $\dfrac{\$0.67}{7\,years} \approx 0.096$ dollars per year. On average, the price of gas increased by about 9.6 cents each year.

Rate of Change

A **rate of change** describes how the output quantity changes in relation to the input quantity. The units on a rate of change are "output units per input units"

Some other examples of rates of change would be quantities like:
- A population of rats increases by 40 rats per week
- A barista earns $9 per hour (dollars per hour)
- A farmer plants 60,000 onions per acre
- A car can drive 27 miles per gallon
- A population of grey whales decreases by 8 whales per year
- The amount of money in your college account decreases by $4,000 per quarter

---

Average Rate of Change

The **average rate of change** between two input values is the total change of the function values (output values) divided by the change in the input values.

$$\text{Average rate of change} = \frac{\text{Change of Output}}{\text{Change of Input}} = \frac{\Delta y}{\Delta x} = \frac{y_2 - y_1}{x_2 - x_1}$$

Example 1

Using the cost-of-gas function from earlier, find the average rate of change between 2007 and 2009

From the table, in 2007 the cost of gas was $2.64. In 2009 the cost was $2.14.

The input (years) has changed by 2. The output has changed by $2.14 - $2.64 = -0.50.

The average rate of change is then $\dfrac{-\$0.50}{2\,years}$ = -0.25 dollars per year

---

Try it Now

1. Using the same cost-of-gas function, find the average rate of change between 2003 and 2008

Notice that in the last example the change of output was *negative* since the output value of the function had decreased. Correspondingly, the average rate of change is negative.

Example 2

Given the function *g(t)* shown here, find the average rate of change on the interval [0, 3].

At $t = 0$, the graph shows $g(0) = 1$
At $t = 3$, the graph shows $g(3) = 4$

The output has changed by 3 while the input has changed by 3, giving an average rate of change of:

$$\frac{4-1}{3-0} = \frac{3}{3} = 1$$

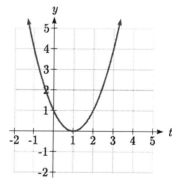

Example 3

On a road trip, after picking up your friend who lives 10 miles away, you decide to record your distance from home over time. Find your average speed over the first 6 hours.

t (hours)	0	1	2	3	4	5	6	7
$D(t)$ (miles)	10	55	90	153	214	240	292	300

Here, your average speed is the average rate of change.
You traveled 282 miles in 6 hours, for an average speed of
$$\frac{292-10}{6-0} = \frac{282}{6} = 47 \text{ miles per hour}$$

We can more formally state the average rate of change calculation using function notation.

Average Rate of Change using Function Notation

Given a function $f(x)$, the average rate of change on the interval $[a, b]$ is

$$\text{Average rate of change} = \frac{\text{Change of Output}}{\text{Change of Input}} = \frac{f(b) - f(a)}{b - a}$$

Example 4

Compute the average rate of change of $f(x) = x^2 - \dfrac{1}{x}$ on the interval $[2, 4]$

We can start by computing the function values at each endpoint of the interval
$$f(2) = 2^2 - \frac{1}{2} = 4 - \frac{1}{2} = \frac{7}{2}$$
$$f(4) = 4^2 - \frac{1}{4} = 16 - \frac{1}{4} = \frac{63}{4}$$

Now computing the average rate of change
$$\text{Average rate of change} = \frac{f(4) - f(2)}{4 - 2} = \frac{\dfrac{63}{4} - \dfrac{7}{2}}{4 - 2} = \frac{\dfrac{49}{4}}{2} = \frac{49}{8}$$

Try it Now

2. Find the average rate of change of $f(x) = x - 2\sqrt{x}$ on the interval $[1, 9]$

Example 5

The magnetic force F, measured in Newtons, between two magnets is related to the distance between the magnets d, in centimeters, by the formula $F(d) = \dfrac{2}{d^2}$. Find the average rate of change of force if the distance between the magnets is increased from 2 cm to 6 cm.

We are computing the average rate of change of $F(d) = \dfrac{2}{d^2}$ on the interval [2, 6].

Average rate of change $= \dfrac{F(6) - F(2)}{6 - 2}$ Evaluating the function

$$\dfrac{F(6) - F(2)}{6 - 2} =$$

$$\dfrac{\dfrac{2}{6^2} - \dfrac{2}{2^2}}{6 - 2}$$ Simplifying

$$\dfrac{\dfrac{2}{36} - \dfrac{2}{4}}{4}$$ Combining the numerator terms

$$\dfrac{\dfrac{-16}{36}}{4}$$ Simplifying further

$$\dfrac{-1}{9}$$ Newtons per centimeter

This tells us the magnetic force decreases, on average, by 1/9 Newtons per centimeter over this interval.

Example 6

Find the average rate of change of $g(t) = t^2 + 3t + 1$ on the interval $[0, a]$. Your answer will be an expression involving a.

Using the average rate of change formula

$$\dfrac{g(a) - g(0)}{a - 0}$$ Evaluating the function

$$\dfrac{(a^2 + 3a + 1) - (0^2 + 3(0) + 1)}{a - 0}$$ Simplifying

$$\dfrac{a^2+3a+1-1}{a}$$ Simplifying further, and factoring

$$\dfrac{a(a+3)}{a}$$ Cancelling the common factor a

$$a+3$$

This result tells us the average rate of change between $t=0$ and any other point $t=a$. For example, on the interval [0, 5], the average rate of change would be 5+3 = 8.

Try it Now

3. Find the average rate of change of $f(x)=x^3+2$ on the interval $[a,a+h]$.

Graphical Behavior of Functions

As part of exploring how functions change, it is interesting to explore the graphical behavior of functions.

Increasing/Decreasing

A function is **increasing** on an interval if the function values increase as the inputs increase. More formally, a function is increasing if $f(b) > f(a)$ for any two input values a and b in the interval with $b>a$. The average rate of change of an increasing function is **positive.**

A function is **decreasing** on an interval if the function values decrease as the inputs increase. More formally, a function is decreasing if $f(b) < f(a)$ for any two input values a and b in the interval with $b>a$. The average rate of change of a decreasing function is **negative.**

Example 7

Given the function $p(t)$ graphed here, on what intervals does the function appear to be increasing?

The function appears to be increasing from $t=1$ to $t=3$, and from $t=4$ on.

In interval notation, we would say the function appears to be increasing on the interval $(1,3)$ and the interval $(4,\infty)$.

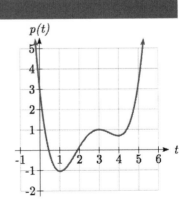

Notice in the last example that we used open intervals (intervals that don't include the endpoints) since the function is neither increasing nor decreasing at $t = 1, 3$, or 4.

Local Extrema
A point where a function changes from increasing to decreasing is called a **local maximum**.
A point where a function changes from decreasing to increasing is called a **local minimum**.
Together, local maxima and minima are called the **local extrema**, or local extreme values, of the function.

Example 8

Using the cost of gasoline function from the beginning of the section, find an interval on which the function appears to be decreasing. Estimate any local extrema using the table.

t	2	3	4	5	6	7	8	9
$C(t)$	1.47	1.69	1.94	2.30	2.51	2.64	3.01	2.14

It appears that the cost of gas increased from $t = 2$ to $t = 8$. It appears the cost of gas decreased from $t = 8$ to $t = 9$, so the function appears to be decreasing on the interval $(8, 9)$.

Since the function appears to change from increasing to decreasing at $t = 8$, there is local maximum at $t = 8$.

Example 9

Use a graph to estimate the local extrema of the function $f(x) = \dfrac{2}{x} + \dfrac{x}{3}$. Use these to determine the intervals on which the function is increasing.

Using technology to graph the function, it appears there is a local minimum somewhere between $x = 2$ and $x = 3$, and a symmetric local maximum somewhere between $x = -3$ and $x = -2$.

Most graphing calculators and graphing utilities can estimate the location of maxima and minima. Below are screen images from two different technologies, showing the estimate for the local maximum and minimum.

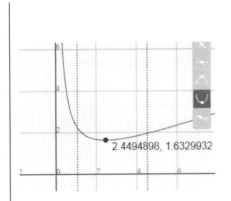 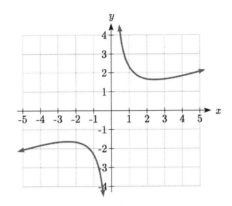

Based on these estimates, the function is increasing on the intervals $(-\infty, -2.449)$ and $(2.449, \infty)$. Notice that while we expect the extrema to be symmetric, the two different technologies agree only up to 4 decimals due to the differing approximation algorithms used by each.

Try it Now

4. Use a graph of the function $f(x) = x^3 - 6x^2 - 15x + 20$ to estimate the local extrema of the function. Use these to determine the intervals on which the function is increasing and decreasing.

Concavity

The total sales, in thousands of dollars, for two companies over 4 weeks are shown.

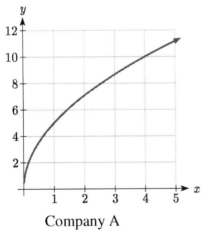 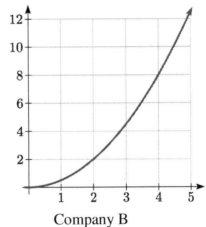

Company A   Company B

As you can see, the sales for each company are increasing, but they are increasing in very different ways. To describe the difference in behavior, we can investigate how the average rate of change varies over different intervals. Using tables of values,

Company A

Week	Sales	Rate of Change
0	0	
		5
1	5	
		2.1
2	7.1	
		1.6
3	8.7	
		1.3
4	10	

Company B

Week	Sales	Rate of Change
0	0	
		0.5
1	0.5	
		1.5
2	2	
		2.5
3	4.5	
		3.5
4	8	

From the tables, we can see that the rate of change for company A is *decreasing*, while the rate of change for company B is *increasing*.

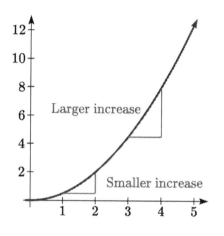 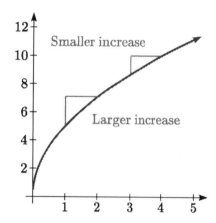

When the rate of change is getting smaller, as with Company A, we say the function is **concave down**. When the rate of change is getting larger, as with Company B, we say the function is **concave up**.

Concavity
A function is **concave up** if the rate of change is increasing.
A function is **concave down** if the rate of change is decreasing.
A point where a function changes from concave up to concave down or vice versa is called an **inflection point**.

Example 10

An object is thrown from the top of a building. The object's height in feet above ground after t seconds is given by the function $h(t) = 144 - 16t^2$ for $0 \le t \le 3$. Describe the concavity of the graph.

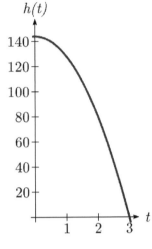

Sketching a graph of the function, we can see that the function is decreasing. We can calculate some rates of change to explore the behavior.

t	$h(t)$	Rate of Change
0	144	
		-16
1	128	
		-48
2	80	
		-80
3	0	

Notice that the rates of change are becoming more negative, so the rates of change are *decreasing*. This means the function is concave down.

Example 11

The value, V, of a car after t years is given in the table below. Is the value increasing or decreasing? Is the function concave up or concave down?

t	0	2	4	6	8
$V(t)$	28000	24342	21162	18397	15994

can compute rates of change to determine concavity.

t	0	2	4	6	8
$V(t)$	28000	24342	21162	18397	15994
Rate of change		-1829	-1590	-1382.5	-1201.5

Since these values are becoming less negative, the rates of change are *increasing*, so this function is concave up.

Try it Now

5. Is the function described in the table below concave up or concave down?

x	0	5	10	15	20
$g(x)$	10000	9000	7000	4000	0

Graphically, concave down functions bend downwards like a frown, and concave up function bend upwards like a smile.

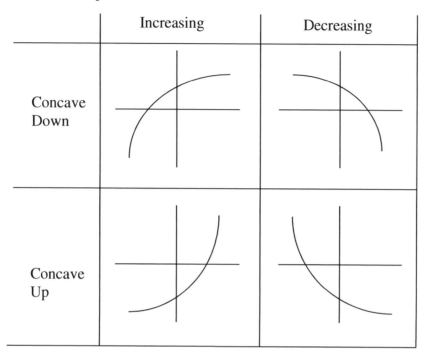

	Increasing	Decreasing
Concave Down		
Concave Up		

Estimate from the graph shown the intervals on which the function is concave down and concave up.

On the far left, the graph is decreasing but concave up, since it is bending upwards. It begins increasing at $x = -2$, but it continues to bend upwards until about $x = -1$.

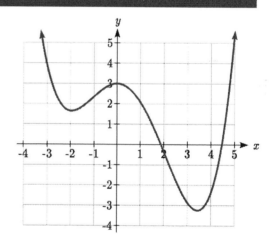

From $x = -1$ the graph starts to bend downward, and continues to do so until about $x = 2$. The graph then begins curving upwards for the remainder of the graph shown.

From this, we can estimate that the graph is concave up on the intervals $(-\infty, -1)$ and $(2, \infty)$, and is concave down on the interval $(-1, 2)$. The graph has inflection points at $x = -1$ and $x = 2$.

Try it Now

6. Using the graph from Try it Now 4, $f(x) = x^3 - 6x^2 - 15x + 20$, estimate the intervals on which the function is concave up and concave down.

Behaviors of the Toolkit Functions

We will now return to our toolkit functions and discuss their graphical behavior.

Function	Increasing/Decreasing	Concavity
Constant Function $f(x) = c$	Neither increasing nor decreasing	Neither concave up nor down
Identity Function $f(x) = x$	Increasing	Neither concave up nor down
Quadratic Function $f(x) = x^2$	Increasing on $(0, \infty)$ Decreasing on $(-\infty, 0)$ Minimum at $x = 0$	Concave up $(-\infty, \infty)$
Cubic Function $f(x) = x^3$	Increasing	Concave down on $(-\infty, 0)$ Concave up on $(0, \infty)$ Inflection point at $(0,0)$
Reciprocal $f(x) = \dfrac{1}{x}$	Decreasing $(-\infty, 0) \cup (0, \infty)$	Concave down on $(-\infty, 0)$ Concave up on $(0, \infty)$

Function	Increasing/Decreasing	Concavity
Reciprocal squared $f(x) = \dfrac{1}{x^2}$	Increasing on $(-\infty, 0)$ Decreasing on $(0, \infty)$	Concave up on $(-\infty, 0) \cup (0, \infty)$
Cube Root $f(x) = \sqrt[3]{x}$	Increasing	Concave down on $(0, \infty)$ Concave up on $(-\infty, 0)$ Inflection point at $(0,0)$
Square Root $f(x) = \sqrt{x}$	Increasing on $(0, \infty)$	Concave down on $(0, \infty)$
Absolute Value $f(x) = |x|$	Increasing on $(0, \infty)$ Decreasing on $(-\infty, 0)$	Neither concave up or down

Important Topics of This Section
Rate of Change
Average Rate of Change
Calculating Average Rate of Change using Function Notation
Increasing/Decreasing
Local Maxima and Minima (Extrema)
Inflection points
Concavity

Try it Now Answers

1. $\dfrac{\$3.01 - \$1.69}{5\,years} = \dfrac{\$1.32}{5\,years} = 0.264$ dollars per year.

2. Average rate of change $= \dfrac{f(9) - f(1)}{9 - 1} = \dfrac{\left(9 - 2\sqrt{9}\right) - \left(1 - 2\sqrt{1}\right)}{9 - 1} = \dfrac{(3) - (-1)}{9 - 1} = \dfrac{4}{8} = \dfrac{1}{2}$

3. $\dfrac{f(a+h) - f(a)}{(a+h) - a} = \dfrac{\left((a+h)^3 + 2\right) - \left(a^3 + 2\right)}{h} = \dfrac{a^3 + 3a^2h + 3ah^2 + h^3 + 2 - a^3 - 2}{h} =$

$\dfrac{3a^2h + 3ah^2 + h^3}{h} = \dfrac{h\left(3a^2 + 3ah + h^2\right)}{h} = 3a^2 + 3ah + h^2$

4. Based on the graph, the local maximum appears to occur at (-1, 28), and the local minimum occurs at (5,-80). The function is increasing on $(-\infty, -1) \cup (5, \infty)$ and decreasing on $(-1, 5)$.

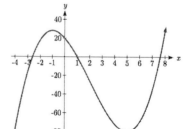

5. Calculating the rates of change, we see the rates of change become *more* negative, so the rates of change are *decreasing*. This function is concave down.

x	0	5	10	15	20
g(x)	10000	9000	7000	4000	0
Rate of change		-1000	-2000	-3000	-4000

6. Looking at the graph, it appears the function is concave down on $(-\infty, 2)$ and concave up on $(2, \infty)$.

Section 1.3 Exercises

1. The table below gives the annual sales (in millions of dollars) of a product. What was the average rate of change of annual sales...
   a) Between 2001 and 2002?   b) Between 2001 and 2004?

year	1998	1999	2000	2001	2002	2003	2004	2005	2006
sales	201	219	233	243	249	251	249	243	233

2. The table below gives the population of a town, in thousands. What was the average rate of change of population...
   a) Between 2002 and 2004?   b) Between 2002 and 2006?

year	2000	2001	2002	2003	2004	2005	2006	2007	2008
population	87	84	83	80	77	76	75	78	81

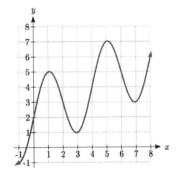

3. Based on the graph shown, estimate the average rate of change from $x = 1$ to $x = 4$.

4. Based on the graph shown, estimate the average rate of change from $x = 2$ to $x = 5$.

Find the average rate of change of each function on the interval specified.

5. $f(x) = x^2$ on [1, 5]

6. $q(x) = x^3$ on [-4, 2]

7. $g(x) = 3x^3 - 1$ on [-3, 3]

8. $h(x) = 5 - 2x^2$ on [-2, 4]

9. $k(t) = 6t^2 + \dfrac{4}{t^3}$ on [-1, 3]

10. $p(t) = \dfrac{t^2 - 4t + 1}{t^2 + 3}$ on [-3, 1]

Find the average rate of change of each function on the interval specified. Your answers will be expressions involving a parameter (*b* or *h*).

11. $f(x) = 4x^2 - 7$ on [1, *b*]

12. $g(x) = 2x^2 - 9$ on [4, *b*]

13. $h(x) = 3x + 4$ on [2, 2+*h*]

14. $k(x) = 4x - 2$ on [3, 3+*h*]

15. $a(t) = \dfrac{1}{t + 4}$ on [9, 9+*h*]

16. $b(x) = \dfrac{1}{x + 3}$ on [1, 1+*h*]

17. $j(x) = 3x^3$ on [1, 1+*h*]

18. $r(t) = 4t^3$ on [2, 2+*h*]

19. $f(x) = 2x^2 + 1$ on [*x*, *x*+*h*]

20. $g(x) = 3x^2 - 2$ on [*x*, *x*+*h*]

For each function graphed, estimate the intervals on which the function is increasing and decreasing.

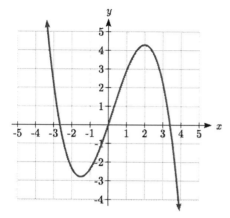

21.

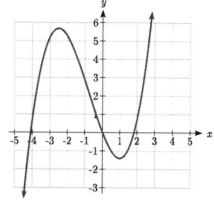

22.

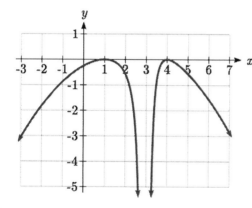

23.

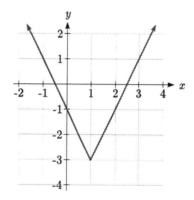

24.

For each table below, select whether the table represents a function that is increasing or decreasing, and whether the function is concave up or concave down.

25.

x	f(x)
1	2
2	4
3	8
4	16
5	32

26.

x	g(x)
1	90
2	80
3	75
4	72
5	70

27.

x	h(x)
1	300
2	290
3	270
4	240
5	200

28.

x	k(x)
1	0
2	15
3	25
4	32
5	35

29.

x	f(x)
1	-10
2	-25
3	-37
4	-47
5	-54

30.

x	g(x)
1	-200
2	-190
3	-160
4	-100
5	0

31.

x	h(x)
1	-100
2	-50
3	-25
4	-10
5	0

32.

x	k(x)
1	-50
2	-100
3	-200
4	-400
5	-900

For each function graphed, estimate the intervals on which the function is concave up and concave down, and the location of any inflection points.

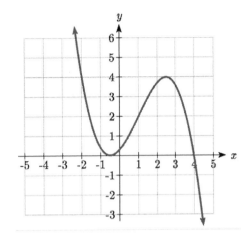

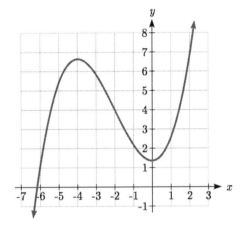

33.   34.

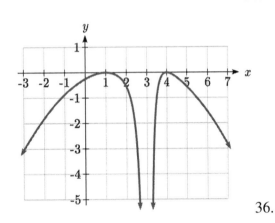

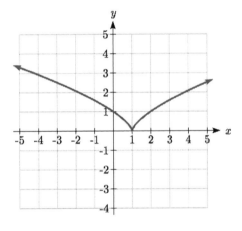

35.   36.

Use a graph to estimate the local extrema and inflection points of each function, and to estimate the intervals on which the function is increasing, decreasing, concave up, and concave down.

37. $f(x) = x^4 - 4x^3 + 5$

38. $h(x) = x^5 + 5x^4 + 10x^3 + 10x^2 - 1$

39. $g(t) = t\sqrt{t+3}$

40. $k(t) = 3t^{2/3} - t$

41. $m(x) = x^4 + 2x^3 - 12x^2 - 10x + 4$

42. $n(x) = x^4 - 8x^3 + 18x^2 - 6x + 2$

Section 1.4 Composition of Functions

Suppose we wanted to calculate how much it costs to heat a house on a particular day of the year. The cost to heat a house will depend on the average daily temperature, and the average daily temperature depends on the particular day of the year. Notice how we have just defined two relationships: The temperature depends on the day, and the cost depends on the temperature. Using descriptive variables, we can notate these two functions.

The first function, $C(T)$, gives the cost C of heating a house when the average daily temperature is T degrees Celsius, and the second, $T(d)$, gives the average daily temperature on day d of the year in some city. If we wanted to determine the cost of heating the house on the 5th day of the year, we could do this by linking our two functions together, an idea called composition of functions. Using the function $T(d)$, we could evaluate $T(5)$ to determine the average daily temperature on the 5th day of the year. We could then use that temperature as the input to the $C(T)$ function to find the cost to heat the house on the 5th day of the year: $C(T(5))$.

Composition of Functions

When the output of one function is used as the input of another, we call the entire operation a **composition of functions**. We write $f(g(x))$, and read this as "f of g of x" or "f composed with g at x".

An alternate notation for composition uses the composition operator: $\circ$

$(f \circ g)(x)$ is read "f of g of x" or "f composed with g at x", just like $f(g(x))$.

Example 1

Suppose $c(s)$ gives the number of calories burned doing s sit-ups, and $s(t)$ gives the number of sit-ups a person can do in t minutes. Interpret $c(s(3))$.

When we are asked to interpret, we are being asked to explain the meaning of the expression in words. The inside expression in the composition is $s(3)$. Since the input to the s function is time, the 3 is representing 3 minutes, and $s(3)$ is the number of sit-ups that can be done in 3 minutes. Taking this output and using it as the input to the $c(s)$ function will gives us the calories that can be burned by the number of sit-ups that can be done in 3 minutes.

Note that it is not important that the same variable be used for the output of the inside function and the input to the outside function. However, it *is* essential that the units on the output of the inside function match the units on the input to the outside function, if the units are specified.

Example 2

Suppose $f(x)$ gives miles that can be driven in x hours, and $g(y)$ gives the gallons of gas used in driving y miles. Which of these expressions is meaningful: $f(g(y))$ or $g(f(x))$?

The expression $g(y)$ takes miles as the input and outputs a number of gallons. The function $f(x)$ is expecting a number of hours as the input; trying to give it a number of gallons as input does not make sense. Remember the units must match, and number of gallons does not match number of hours, so the expression $f(g(y))$ is meaningless.

The expression $f(x)$ takes hours as input and outputs a number of miles driven. The function $g(y)$ is expecting a number of miles as the input, so giving the output of the $f(x)$ function (miles driven) as an input value for $g(y)$, where gallons of gas depends on miles driven, does make sense. The expression $g(f(x))$ makes sense, and will give the number of gallons of gas used, g, driving a certain number of miles, $f(x)$, in x hours.

Try it Now

1. In a department store you see a sign that says 50% off clearance merchandise, so final cost C depends on the clearance price, p, according to the function $C(p)$. Clearance price, p, depends on the original discount, d, given to the clearance item, $p(d)$. Interpret $C(p(d))$.

Composition of Functions using Tables and Graphs

When working with functions given as tables and graphs, we can look up values for the functions using a provided table or graph, as discussed in section 1.1. We start evaluation from the provided input, and first evaluate the inside function. We can then use the output of the inside function as the input to the outside function. To remember this, always work from the inside out.

Example 3

Using the tables below, evaluate $f(g(3))$ and $g(f(4))$

x	$f(x)$
1	6
2	8
3	3
4	1

x	$g(x)$
1	3
2	5
3	2
4	7

To evaluate $f(g(3))$, we start from the inside with the value 3. We then evaluate the inside expression $g(3)$ using the table that defines the function g: $g(3) = 2$.

We can then use that result as the input to the *f* function, so $g(3)$ is replaced by the equivalent value 2 and we can evaluate $f(2)$. Then using the table that defines the function *f*, we find that $f(2) = 8$.

$f(g(3)) = f(2) = 8$.

To evaluate $g(f(4))$, we first evaluate the inside expression $f(4)$ using the first table: $f(4) = 1$. Then using the table for *g* we can evaluate:

$g(f(4)) = g(1) = 3$.

Try it Now

2. Using the tables from the example above, evaluate $f(g(1))$ and $g(f(3))$.

Example 4

Using the graphs below, evaluate $f(g(1))$.

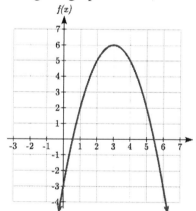
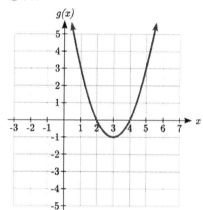

To evaluate $f(g(1))$, we again start with the inside evaluation. We evaluate $g(1)$ using the graph of the *g(x)* function, finding the input of 1 on the horizontal axis and finding the output value of the graph at that input. Here, $g(1) = 3$.

Using this value as the input to the *f* function, $f(g(1)) = f(3)$. We can then evaluate this by looking to the graph of the *f(x)* function, finding the input of 3 on the horizontal axis, and reading the output value of the graph at this input.

$f(3) = 6$, so $f(g(1)) = 6$.

Try it Now

3. Using the graphs from the previous example, evaluate $g(f(2))$.

Composition using Formulas

When evaluating a composition of functions where we have either created or been given formulas, the concept of working from the inside out remains the same. First, we evaluate the inside function using the input value provided, then use the resulting output as the input to the outside function.

Example 5

Given $f(t) = t^2 - t$ and $h(x) = 3x + 2$, evaluate $f(h(1))$.

Since the inside evaluation is $h(1)$ we start by evaluating the $h(x)$ function at 1:
$h(1) = 3(1) + 2 = 5$

Then $f(h(1)) = f(5)$, so we evaluate the $f(t)$ function at an input of 5:
$f(h(1)) = f(5) = 5^2 - 5 = 20$

Try it Now
4. Using the functions from the example above, evaluate $h(f(-2))$.

While we can compose the functions as above for each individual input value, sometimes it would be really helpful to find a single formula which will calculate the result of a composition $f(g(x))$. To do this, we will extend our idea of function evaluation. Recall that when we evaluate a function like $f(t) = t^2 - t$, we put whatever value is inside the parentheses after the function name into the formula wherever we see the input variable.

Example 6

Given $f(t) = t^2 - t$, evaluate $f(3)$ and $f(-2)$.

$f(3) = 3^2 - 3$
$f(-2) = (-2)^2 - (-2)$

We could simplify the results above if we wanted to
$f(3) = 3^2 - 3 = 9 - 3 = 6$
$f(-2) = (-2)^2 - (-2) = 4 + 2 = 6$

We are not limited, however, to using a numerical value as the input to the function. We can put anything into the function: a value, a different variable, or even an algebraic expression, provided we use the input expression everywhere we see the input variable.

Example 7

Using the function from the previous example, evaluate *f(a)*.

This means that the input value for *t* is some unknown quantity *a*. As before, we evaluate by replacing the input variable *t* with the input quantity, in this case *a*.

$$f(a) = a^2 - a$$

The same idea can then be applied to expressions more complicated than a single letter.

Example 8

Using the same *f(t)* function from above, evaluate $f(x+2)$.

Everywhere in the formula for *f* where there was a *t*, we would replace it with the input $(x+2)$. Since in the original formula the input *t* was squared in the first term, the entire input $x+2$ needs to be squared when we substitute, so we need to use grouping parentheses. To avoid problems, it is advisable to always use parentheses around inputs.

$$f(x+2) = (x+2)^2 - (x+2)$$

We could simplify this expression further to $f(x+2) = x^2 + 3x + 2$ if we wanted to:

$f(x+2) = (x+2)(x+2) - (x+2)$ Use the "FOIL" technique (first, outside, inside, last)

$f(x+2) = x^2 + 2x + 2x + 4 - (x+2)$ distribute the negative sign

$f(x+2) = x^2 + 2x + 2x + 4 - x - 2$ combine like terms

$f(x+2) = x^2 + 3x + 2$

Example 9

Using the same function, evaluate $f(t^3)$.

Note that in this example, the same variable is used in the input expression and as the input variable of the function. This doesn't matter – we still replace the original input *t* in the formula with the new input expression, t^3.

$$f(t^3) = (t^3)^2 - (t^3) = t^6 - t^3$$

5. Given $g(x) = 3x - \sqrt{x}$, evaluate $g(t - 2)$.

This now allows us to find an expression for a composition of functions. If we want to find a formula for $f(g(x))$, we can start by writing out the formula for $g(x)$. We can then evaluate the function $f(x)$ at that expression, as in the examples above.

Example 10

Let $f(x) = x^2$ and $g(x) = \dfrac{1}{x} - 2x$, find $f(g(x))$ and $g(f(x))$.

To find $f(g(x))$, we start by evaluating the inside, writing out the formula for $g(x)$.

$$g(x) = \dfrac{1}{x} - 2x$$

We then use the expression $\left(\dfrac{1}{x} - 2x \right)$ as input for the function f.

$$f(g(x)) = f\left(\dfrac{1}{x} - 2x \right)$$

We then evaluate the function $f(x)$ using the formula for $g(x)$ as the input.

Since $f(x) = x^2$, $f\left(\dfrac{1}{x} - 2x \right) = \left(\dfrac{1}{x} - 2x \right)^2$

This gives us the formula for the composition: $f(g(x)) = \left(\dfrac{1}{x} - 2x \right)^2$.

Likewise, to find $g(f(x))$, we evaluate the inside, writing out the formula for $f(x)$

$$g(f(x)) = g\left(x^2 \right)$$

Now we evaluate the function $g(x)$ using x^2 as the input.

$$g(f(x)) = \dfrac{1}{x^2} - 2x^2$$

6. Let $f(x) = x^3 + 3x$ and $g(x) = \sqrt{x}$, find $f(g(x))$ and $g(f(x))$.

Example 11

A city manager determines that the tax revenue, R, in millions of dollars collected on a population of p thousand people is given by the formula $R(p) = 0.03p + \sqrt{p}$, and that the city's population, in thousands, is predicted to follow the formula $p(t) = 60 + 2t + 0.3t^2$, where t is measured in years after 2010. Find a formula for the tax revenue as a function of the year.

Since we want tax revenue as a function of the year, we want year to be our initial input, and revenue to be our final output. To find revenue, we will first have to predict the city population, and then use that result as the input to the tax function. So we need to find $R(p(t))$. Evaluating this,

$$R(p(t)) = R(60 + 2t + 0.3t^2) = 0.03(60 + 2t + 0.3t^2) + \sqrt{60 + 2t + 0.3t^2}$$

This composition gives us a single formula which can be used to predict the tax revenue during a given year, without needing to find the intermediary population value.

For example, to predict the tax revenue in 2017, when $t = 7$ (because t is measured in years after 2010),

$$R(p(7)) = 0.03(60 + 2(7) + 0.3(7)^2) + \sqrt{60 + 2(7) + 0.3(7)^2} \approx 12.079 \text{ million dollars}$$

Domain of Compositions

When we think about the domain of a composition $h(x) = f(g(x))$, we must consider both the domain of the inner function and the domain of the composition itself. While it is tempting to only look at the resulting composite function, if the inner function were undefined at a value of x, the composition would not be possible.

Example 12

Let $f(x) = \dfrac{1}{x^2 - 1}$ and $g(x) = \sqrt{x - 2}$. Find the domain of $f(g(x))$.

Since we want to avoid the square root of negative numbers, the domain of $g(x)$ is the set of values where $x - 2 \geq 0$. The domain is $x \geq 2$.

The composition is $f(g(x)) = \dfrac{1}{\left(\sqrt{x-2}\right)^2 - 1} = \dfrac{1}{(x-2) - 1} = \dfrac{1}{x - 3}$.

The composition is undefined when $x = 3$, so that value must also be excluded from the domain. Notice that the composition doesn't involve a square root, but we still have to consider the domain limitation from the inside function.

Combining the two restrictions, the domain is all values of x greater than or equal to 2, except $x = 3$.

In inequalities, the domain is: $2 \le x < 3$ or $x > 3$.
In interval notation, the domain is: $[2,3) \cup (3, \infty)$.

Try it Now

7. Let $f(x) = \dfrac{1}{x-2}$ and $g(x) = \dfrac{1}{x}$. Find the domain of $f(g(x))$.

Decomposing Functions

In some cases, it is desirable to decompose a function – to write it as a composition of two simpler functions.

Example 13

Write $f(x) = 3 + \sqrt{5 - x^2}$ as the composition of two functions.

We are looking for two functions, g and h, so $f(x) = g(h(x))$. To do this, we look for a function inside a function in the formula for $f(x)$. As one possibility, we might notice that $5 - x^2$ is the inside of the square root. We could then decompose the function as:

$h(x) = 5 - x^2$

$g(x) = 3 + \sqrt{x}$

We can check our answer by recomposing the functions:

$g(h(x)) = g\left(5 - x^2\right) = 3 + \sqrt{5 - x^2}$

Note that this is not the only solution to the problem. Another non-trivial decomposition would be $h(x) = x^2$ and $g(x) = 3 + \sqrt{5 - x}$

Important Topics of this Section
Definition of Composition of Functions
Compositions using:
Words
Tables
Graphs
Equations
Domain of Compositions
Decomposition of Functions

Try it Now Answers

1. The final cost, C, depends on the clearance price, p, which is based on the original discount, d. (Or the original discount d, determines the clearance price and the final cost is half of the clearance price.)

2. $f(g(1)) = f(3) = 3$ and $g(f(3)) = g(3) = 2$

3. $g(f(2)) = g(5) = 3$

4. $h(f(-2)) = h(6) = 20$ *did you remember to insert your input values using parentheses?*

5. $g(t-2) = 3(t-2) - \sqrt{(t-2)}$

6. $f(g(x)) = f\left(\sqrt{x}\right) = \left(\sqrt{x}\right)^3 + 3\left(\sqrt{x}\right)$

    $g(f(x)) = g\left(x^3 + 3x\right) = \sqrt{\left(x^3 + 3x\right)}$

7. $g(x) = \dfrac{1}{x}$ is undefined at $x = 0$.

The composition, $f\left(g(x)\right) = f\left(\dfrac{1}{x}\right) = \dfrac{1}{\dfrac{1}{x} - 2} = \dfrac{1}{\dfrac{1}{x} - \dfrac{2x}{x}} = \dfrac{1}{\dfrac{1-2x}{x}} = \dfrac{x}{1-2x}$ is undefined

when $1 - 2x = 0$, when $x = \dfrac{1}{2}$.

Restricting these two values, the domain is $(-\infty, 0) \cup \left(0, \dfrac{1}{2}\right) \cup \left(\dfrac{1}{2}, \infty\right)$.

Section 1.4 Exercises

Given each pair of functions, calculate $f(g(0))$ and $g(f(0))$.

1. $f(x) = 4x + 8$, $g(x) = 7 - x^2$
2. $f(x) = 5x + 7$, $g(x) = 4 - 2x^2$
3. $f(x) = \sqrt{x+4}$, $g(x) = 12 - x^3$
4. $f(x) = \dfrac{1}{x+2}$, $g(x) = 4x + 3$

Use the table of values to evaluate each expression

5. $f(g(8))$
6. $f(g(5))$
7. $g(f(5))$
8. $g(f(3))$
9. $f(f(4))$
10. $f(f(1))$
11. $g(g(2))$
12. $g(g(6))$

x	$f(x)$	$g(x)$
0	7	9
1	6	5
2	5	6
3	8	2
4	4	1
5	0	8
6	2	7
7	1	3
8	9	4
9	3	0

Use the graphs to evaluate the expressions below.

13. $f(g(3))$
14. $f(g(1))$
15. $g(f(1))$
16. $g(f(0))$
17. $f(f(5))$
18. $f(f(4))$
19. $g(g(2))$
20. $g(g(0))$

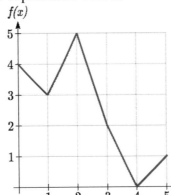
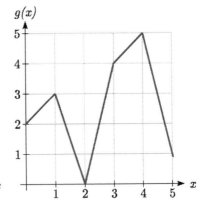

For each pair of functions, find $f(g(x))$ and $g(f(x))$. Simplify your answers.

21. $f(x) = \dfrac{1}{x-6}$, $g(x) = \dfrac{7}{x} + 6$
22. $f(x) = \dfrac{1}{x-4}$, $g(x) = \dfrac{2}{x} + 4$

23. $f(x) = x^2 + 1$, $g(x) = \sqrt{x+2}$
24. $f(x) = \sqrt{x} + 2$, $g(x) = x^2 + 3$

25. $f(x) = |x|$, $g(x) = 5x + 1$
26. $f(x) = \sqrt[3]{x}$, $g(x) = \dfrac{x+1}{x^3}$

27. If $f(x)=x^4+6$, $g(x)=x-6$ and $h(x)=\sqrt{x}$, find $f(g(h(x)))$

28. If $f(x)=x^2+1$, $g(x)=\dfrac{1}{x}$ and $h(x)=x+3$, find $f(g(h(x)))$

29. The function $D(p)$ gives the number of items that will be demanded when the price is p. The production cost, $C(x)$ is the cost of producing x items. To determine the cost of production when the price is $6, you would do which of the following:
   a. Evaluate $D(C(6))$ b. Evaluate $C(D(6))$
   c. Solve $D(C(x))=6$ d. Solve $C(D(p))=6$

30. The function $A(d)$ gives the pain level on a scale of 0-10 experienced by a patient with d milligrams of a pain reduction drug in their system. The milligrams of drug in the patient's system after t minutes is modeled by $m(t)$. To determine when the patient will be at a pain level of 4, you would need to:
   a. Evaluate $A\big(m(4)\big)$ b. Evaluate $m\big(A(4)\big)$
   c. Solve $A\big(m(t)\big)=4$ d. Solve $m\big(A(d)\big)=4$

31. The radius r, in inches, of a spherical balloon is related to the volume, V, by $r(V)=\sqrt[3]{\dfrac{3V}{4\pi}}$. Air is pumped into the balloon, so the volume after t seconds is given by $V(t)=10+20t$.
   a. Find the composite function $r\big(V(t)\big)$
   b. Find the radius after 20 seconds

32. The number of bacteria in a refrigerated food product is given by $N(T)=23T^2-56T+1$, $3<T<33$, where T is the temperature of the food. When the food is removed from the refrigerator, the temperature is given by $T(t)=5t+1.5$, where t is the time in hours.
   a. Find the composite function $N\big(T(t)\big)$
   b. Find the bacteria count after 4 hours

33. Given $p(x)=\dfrac{1}{\sqrt{x}}$ and $m(x)=x^2-4$, find the domain of $m(p(x))$.

34. Given $p(x)=\dfrac{1}{\sqrt{x}}$ and $m(x)=9-x^2$, find the domain of $m(p(x))$.

35. Given $f(x)=\dfrac{1}{x+3}$ and $g(x)=\dfrac{2}{x-1}$, find the domain of $f\big(g(x)\big)$.

36. Given $f(x)=\dfrac{x}{x+1}$ and $g(x)=\dfrac{4}{x}$, find the domain of $f\big(g(x)\big)$.

37. Given $f(x)=\sqrt{x-2}$ and $g(x)=\dfrac{2}{x^2-3}$, find the domain of $g(f(x))$.

38. Given $f(x)=\sqrt{4-x}$ and $g(x)=\dfrac{1}{x^2-2}$, find the domain of $g(f(x))$.

Find functions $f(x)$ and $g(x)$ so the given function can be expressed as $h(x)=f\big(g(x)\big)$.

39. $h(x)=(x+2)^2$

40. $h(x)=(x-5)^3$

41. $h(x)=\dfrac{3}{x-5}$

42. $h(x)=\dfrac{4}{(x+2)^2}$

43. $h(x)=3+\sqrt{x-2}$

44. $h(x)=4+\sqrt[3]{x}$

45. Let $f(x)$ be a linear function, with form $f(x)=ax+b$ for constants a and b. [UW]

   a. Show that $f\big(f(x)\big)$ is a linear function

   b. Find a function $g(x)$ such that $g\big(g(x)\big)=6x-8$

46. Let $f(x)=\dfrac{1}{2}x+3$ [UW]

   a. Sketch the graphs of $f(x),f\big(f(x)\big),f\big(f\big(f(x)\big)\big)$ on the interval $-2\le x\le 10$.

   b. Your graphs should all intersect at the point (6, 6). The value x = 6 is called a fixed point of the function f(x)since $f(6)=6$; that is, 6 is fixed - it doesn't move when f is applied to it. Give an explanation for why 6 is a fixed point for any function $f(f(f(...f(x)...)))$.

   c. Linear functions (with the exception of $f(x)=x$) can have at most one fixed point. Quadratic functions can have at most two. Find the fixed points of the function $g(x)=x^2-2$.

   d. Give a quadratic function whose fixed points are x = -2 and x = 3.

47. A car leaves Seattle heading east. The speed of the car in mph after m minutes is given by the function $C(m) = \dfrac{70m^2}{10 + m^2}$.   [UW]

    a. Find a function $m = f(s)$ that converts seconds s into minutes m. Write out the formula for the new function $C(f(s))$; what does this function calculate?

    b. Find a function $m = g(h)$ that converts hours h into minutes m. Write out the formula for the new function $C(g(h))$; what does this function calculate?

    c. Find a function $z = v(s)$ that converts mph s into ft/sec z. Write out the formula for the new function $v(C(m))$; what does this function calculate?

Section 1.5 Transformation of Functions

Often when given a problem, we try to model the scenario using mathematics in the form of words, tables, graphs and equations in order to explain or solve it. When building models, it is often helpful to build off of existing formulas or models. Knowing the basic graphs of your tool-kit functions can help you solve problems by being able to model new behavior by adapting something you already know. Unfortunately, the models and existing formulas we know are not always exactly the same as the ones presented in the problems we face.

Fortunately, there are systematic ways to shift, stretch, compress, flip and combine functions to help them become better models for the problems we are trying to solve. We can transform what we already know into what we need, hence the name, "Transformation of functions." When we have a story problem, formula, graph, or table, we can then transform that function in a variety of ways to form new functions.

Shifts

Example 1

To regulate temperature in a green building, air flow vents near the roof open and close throughout the day to allow warm air to escape. The graph below shows the open vents V (in square feet) throughout the day, t in hours after midnight. During the summer, the facilities staff decides to try to better regulate temperature by increasing the amount of open vents by 20 square feet throughout the day. Sketch a graph of this new function.

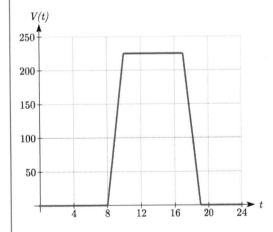

We can sketch a graph of this new function by adding 20 to each of the output values of the original function. This will have the effect of shifting the graph up.

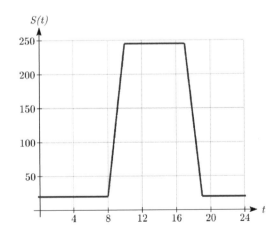

Notice that in the second graph, for each input value, the output value has increased by twenty, so if we call the new function $S(t)$, we could write $S(t) = V(t) + 20$.

Note that this notation tells us that for any value of t, $S(t)$ can be found by evaluating the V function at the same input, then adding twenty to the result.
This defines S as a transformation of the function V, in this case a vertical shift up 20 units.

Notice that with a vertical shift the input values stay the same and only the output values change.

Vertical Shift

Given a function $f(x)$, if we define a new function $g(x)$ as

$g(x) = f(x) + k$, where k is a constant

then $g(x)$ is a **vertical shift** of the function $f(x)$, where all the output values have been increased by k.

If k is positive, then the graph will shift up

If k is negative, then the graph will shift down

Example 2

A function $f(x)$ is given as a table below. Create a table for the function $g(x) = f(x) - 3$

x	2	4	6	8
$f(x)$	1	3	7	11

The formula $g(x) = f(x) - 3$ tells us that we can find the output values of the g function by subtracting 3 from the output values of the f function. For example,

$f(2) = 1$ is found from the given table

$g(x) = f(x) - 3$ is our given transformation

$g(2) = f(2) - 3 = 1 - 3 = -2$

Subtracting 3 from each $f(x)$ value, we can complete a table of values for $g(x)$

x	2	4	6	8
$g(x)$	-2	0	4	8

As with the earlier vertical shift, notice the input values stay the same and only the output values change.

1. The function $h(t) = -4.9t^2 + 30t$ gives the height h of a ball (in meters) thrown upwards from the ground after t seconds. Suppose the ball was instead thrown from the top of a 10 meter building. Relate this new height function $b(t)$ to $h(t)$, then find a formula for $b(t)$.

The vertical shift is a change to the output, or outside, of the function. We will now look at how changes to input, on the inside of the function, change its graph and meaning.

Example 3

Returning to our building air flow example from the beginning of the section, suppose that in Fall, the facilities staff decides that the original venting plan starts too late, and they want to move the entire venting program to start two hours earlier. Sketch a graph of the new function.

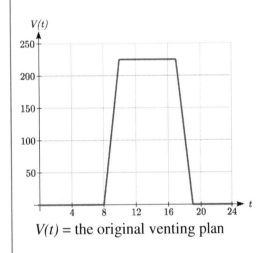

$V(t)$ = the original venting plan

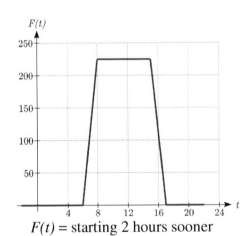

$F(t)$ = starting 2 hours sooner

In the new graph, which we can call $F(t)$, at each time, the air flow is the same as the original function $V(t)$ was two hours later. For example, in the original function V, the air flow starts to change at 8am, while for the function $F(t)$ the air flow starts to change at 6am. The comparable function values are $V(8) = F(6)$.

Notice also that the vents first opened to 220 sq. ft. at 10 a.m. under the original plan, while under the new plan the vents reach 220 sq. ft. at 8 a.m., so $V(10) = F(8)$.

In both cases we see that since $F(t)$ starts 2 hours sooner, the same output values are reached when, $F(t) = V(t + 2)$

Note that $V(t + 2)$ had the effect of shifting the graph to the *left*.

Horizontal changes or "inside changes" affect the domain of a function (the input) instead of the range and often seem counterintuitive. The new function $F(t)$ uses the same outputs as $V(t)$, but matches those outputs to inputs two hours earlier than those of $V(t)$. Said another way, we must add 2 hours to the input of V to find the corresponding output for F: $F(t) = V(t+2)$.

Horizontal Shift

Given a function $f(x)$, if we define a new function $g(x)$ as

$g(x) = f(x+k)$, where k is a constant

then $g(x)$ is a **horizontal shift** of the function $f(x)$

If k is positive, then the graph will shift left

If k is negative, then the graph will shift right

Example 4

A function $f(x)$ is given as a table below. Create a table for the function $g(x) = f(x-3)$

x	2	4	6	8
$f(x)$	1	3	7	11

The formula $g(x) = f(x-3)$ tells us that the output values of g are the same as the output value of f with an input value three smaller. For example, we know that $f(2) = 1$. To get the same output from the g function, we will need an input value that is 3 *larger*. We input a value that is three larger for $g(x)$ because the function takes three away before evaluating the function f.

$g(5) = f(5-3) = f(2) = 1$

x	5	7	9	11
$g(x)$	1	3	7	11

The result is that the function $g(x)$ has been shifted to the right by 3. Notice the output values for $g(x)$ remain the same as the output values for $f(x)$ in the chart, but the corresponding input values, x, have shifted to the right by 3: 2 shifted to 5, 4 shifted to 7, 6 shifted to 9 and 8 shifted to 11.

Example 5

The graph shown is a transformation of the toolkit function $f(x) = x^2$. Relate this new function $g(x)$ to $f(x)$, and then find a formula for $g(x)$.

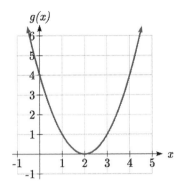

Notice that the graph looks almost identical in shape to the $f(x) = x^2$ function, but the x values are shifted to the right two units. The vertex used to be at $(0, 0)$ but now the vertex is at $(2, 0)$. The graph is the basic quadratic function shifted two to the right, so
$g(x) = f(x-2)$

Notice how we must input the value $x = 2$, to get the output value $y = 0$; the x values must be two units larger, because of the shift to the right by 2 units.

We can then use the definition of the $f(x)$ function to write a formula for $g(x)$ by evaluating $f(x-2)$:
Since $f(x) = x^2$ and $g(x) = f(x-2)$
$g(x) = f(x-2) = (x-2)^2$

If you find yourself having trouble determining whether the shift is +2 or -2, it might help to consider a single point on the graph. For a quadratic, looking at the bottommost point is convenient. In the original function, $f(0) = 0$. In our shifted function, $g(2) = 0$. To obtain the output value of 0 from the f function, we need to decide whether a +2 or -2 will work to satisfy $g(2) = f(2 ? 2) = f(0) = 0$. For this to work, we will need to subtract 2 from our input values.

When thinking about horizontal and vertical shifts, it is good to keep in mind that vertical shifts are affecting the output values of the function, while horizontal shifts are affecting the input values of the function.

Example 6

The function $G(m)$ gives the number of gallons of gas required to drive m miles. Interpret $G(m) + 10$ and $G(m+10)$.

$G(m) + 10$ is adding 10 to the output, gallons. This is 10 gallons of gas more than is required to drive m miles. So, this is the gas required to drive m miles, plus another 10 gallons of gas.

$G(m+10)$ is adding 10 to the input, miles. This is the number of gallons of gas required to drive 10 miles more than m miles.

Try it Now

2. Given the function $f(x) = \sqrt{x}$ graph the original function $f(x)$ and the transformation $g(x) = f(x+2)$.
   a. Is this a horizontal or a vertical change?
   b. Which way is the graph shifted and by how many units?
   c. Graph $f(x)$ and $g(x)$ on the same axes.

Now that we have two transformations, we can combine them together.

Remember:
Vertical Shifts are outside changes that affect the output (vertical) axis values shifting the transformed function up or down.

Horizontal Shifts are inside changes that affect the input (horizontal) axis values shifting the transformed function left or right.

Example 7

Given $f(x) = |x|$, sketch a graph of $h(x) = f(x+1) - 3$.

The function f is our toolkit absolute value function. We know that this graph has a V shape, with the point at the origin. The graph of h has transformed f in two ways: $f(x+1)$ is a change on the inside of the function, giving a horizontal shift left by 1, then the subtraction by 3 in $f(x+1) - 3$ is a change to the outside of the function, giving a vertical shift down by 3. Transforming the graph gives

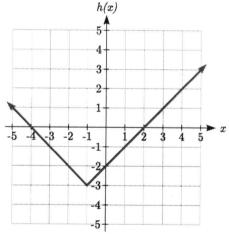

We could also find a formula for this transformation by evaluating the expression for $h(x)$:

$h(x) = f(x+1) - 3$

$h(x) = |x+1| - 3$

Example 8

Write a formula for the graph shown, a transformation of the toolkit square root function.

The graph of the toolkit function starts at the origin, so this graph has been shifted 1 to the right, and up 2. In function notation, we could write that as $h(x) = f(x-1) + 2$. Using the formula for the square root function we can write

$$h(x) = \sqrt{x-1} + 2$$

Note that this transformation has changed the domain and range of the function. This new graph has domain $[1, \infty)$ and range $[2, \infty)$.

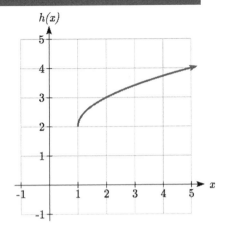

Reflections

Another transformation that can be applied to a function is a reflection over the horizontal or vertical axis.

Example 9

Reflect the graph of $s(t) = \sqrt{t}$ both vertically and horizontally.

Reflecting the graph vertically, each output value will be reflected over the horizontal t axis:

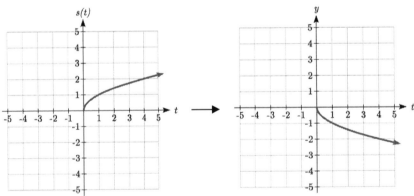

Since each output value is the opposite of the original output value, we can write

$$V(t) = -s(t)$$

$$V(t) = -\sqrt{t}$$

Notice this is an outside change or vertical change that affects the output $s(t)$ values so the negative sign belongs outside of the function.

Reflecting horizontally, each input value will be reflected over the vertical axis.

Since each input value is the opposite of the original input value, we can write

$$H(t) = s(-t)$$

$$H(t) = \sqrt{-t}$$

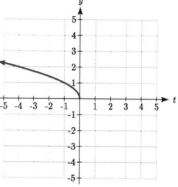

Notice this is an inside change or horizontal change that affects the input values so the negative sign is on the inside of the function.

Note that these transformations can affect the domain and range of the functions. While the original square root function has domain $[0, \infty)$ and range $[0, \infty)$, the vertical reflection gives the $V(t)$ function the range $(-\infty, 0]$, and the horizontal reflection gives the $H(t)$ function the domain $(-\infty, 0]$.

Reflections

Given a function $f(x)$, if we define a new function $g(x)$ as

$$g(x) = -f(x),$$

then $g(x)$ is a **vertical reflection** of the function $f(x)$, sometimes called a reflection about the x-axis

If we define a new function $g(x)$ as

$$g(x) = f(-x),$$

then $g(x)$ is a **horizontal reflection** of the function $f(x)$, sometimes called a reflection about the y-axis

---

Example 10

A function $f(x)$ is given as a table below. Create a table for the function $g(x) = -f(x)$ and $h(x) = f(-x)$

x	2	4	6	8
$f(x)$	1	3	7	11

For $g(x)$, this is a vertical reflection, so the x values stay the same and each output value will be the opposite of the original output value

For $h(x)$, this is a horizontal reflection, and each input value will be the opposite of the original input value and the $h(x)$ values stay the same as the $f(x)$ values:

x	-2	-4	-6	-8
$h(x)$	1	3	7	11

Example 11

A common model for learning has an equation similar to $k(t) = -2^{-t} + 1$, where k is the percentage of mastery that can be achieved after t practice sessions. This is a transformation of the function $f(t) = 2^t$ shown here. Sketch a graph of $k(t)$.

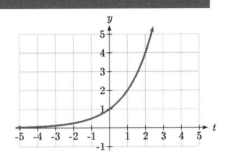

This equation combines three transformations into one equation.

A horizontal reflection: $\qquad f(-t) = 2^{-t}$ $\qquad$ combined with

A vertical reflection: $\qquad -f(-t) = -2^{-t}$ $\qquad$ combined with

A vertical shift up 1: $\qquad -f(-t) + 1 = -2^{-t} + 1$

We can sketch a graph by applying these transformations one at a time to the original function:

The original graph $\qquad$ Horizontally reflected $\qquad$ Then vertically reflected

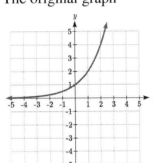 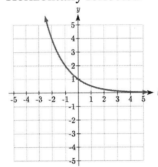 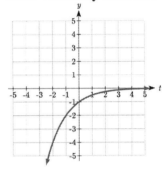

Then, after shifting up 1, we get the final graph.

$k(t) = -f(-t) + 1 = -2^{-t} + 1$.

Note: As a model for learning, this function would be limited to a domain of $t \geq 0$, with corresponding range $[0, 1)$.

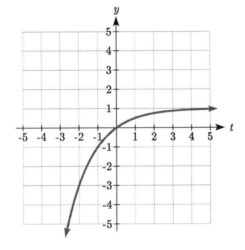

Try it Now

3. Given the toolkit function $f(x) = x^2$, graph $g(x) = -f(x)$ and $h(x) = f(-x)$.
   Do you notice anything surprising?

Some functions exhibit symmetry, in which reflections result in the original graph. For example, reflecting the toolkit functions $f(x) = x^2$ or $f(x) = |x|$ about the y-axis will result in the original graph. We call these types of graphs symmetric about the y-axis.

Likewise, if the graphs of $f(x) = x^3$ or $f(x) = \dfrac{1}{x}$ were reflected over both axes, the result would be the original graph:

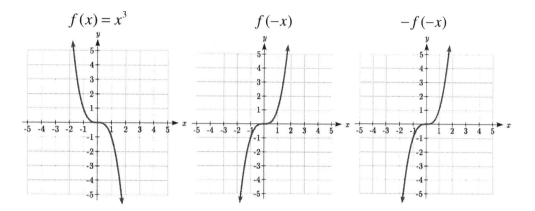

We call these graphs symmetric about the origin.

Even and Odd Functions
A function is called an **even function** if $f(x) = f(-x)$ The graph of an even function is symmetric about the vertical axis A function is called an **odd function** if $f(x) = -f(-x)$ The graph of an odd function is symmetric about the origin

Note: A function can be neither even nor odd if it does not exhibit either symmetry. For example, the $f(x) = 2^x$ function is neither even nor odd.

Example 12

Is the function $f(x) = x^3 + 2x$ even, odd, or neither?

Without looking at a graph, we can determine this by finding formulas for the reflections, and seeing if they return us to the original function:

$$f(-x) = (-x)^3 + 2(-x) = -x^3 - 2x$$

This does not return us to the original function, so this function is not even.

We can now try also applying a horizontal reflection:

$$-f(-x) = -\left(-x^3 - 2x\right) = x^3 + 2x$$

Since $-f(-x) = f(x)$, this is an odd function.

Stretches and Compressions

With shifts, we saw the effect of adding or subtracting to the inputs or outputs of a function. We now explore the effects of multiplying the inputs or outputs.

Remember, we can transform the inside (input values) of a function or we can transform the outside (output values) of a function. Each change has a specific effect that can be seen graphically.

Example 13

A function $P(t)$ models the growth of a population of fruit flies. The growth is shown in the graph. A scientist is comparing this to another population, Q, that grows the same way, but starts twice as large. Sketch a graph of this population.

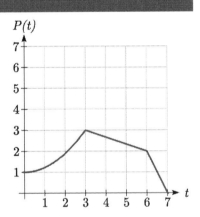

Since the population is always twice as large, the new population's output values are always twice the original function output values. Graphically, this would look like the second graph shown.

Symbolically, $Q(t) = 2P(t)$

This means that for any input t, the value of the Q function is twice the value of the P function. Notice the effect on the graph is a vertical stretching of the graph, where every point doubles its distance from the horizontal axis. The input values, t, stay the same while the output values are twice as large as before.

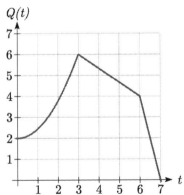

Vertical Stretch/Compression

Given a function *f(x)*, if we define a new function *g(x)* as

$g(x) = kf(x)$, where *k* is a constant

then *g(x)* is a **vertical stretch or compression** of the function *f(x)*.

If $k > 1$, then the graph will be stretched

If $0 < k < 1$, then the graph will be compressed

If $k < 0$, then there will be combination of a vertical stretch or compression with a vertical reflection

Example 14

A function *f(x)* is given as a table below. Create a table for the function $g(x) = \dfrac{1}{2} f(x)$

x	2	4	6	8
f(x)	1	3	7	11

The formula $g(x) = \dfrac{1}{2} f(x)$ tells us that the output values of *g* are half of the output values of *f* with the same inputs. For example, we know that $f(4) = 3$. Then

$$g(4) = \frac{1}{2} f(4) = \frac{1}{2}(3) = \frac{3}{2}$$

x	2	4	6	8
g(x)	1/2	3/2	7/2	11/2

The result is that the function *g(x)* has been compressed vertically by ½. Each output value has been cut in half, so the graph would now be half the original height.

Example 15

The graph shown is a transformation of the toolkit function $f(x) = x^3$. Relate this new function *g(x)* to *f(x)*, then find a formula for *g(x)*.

When trying to determine a vertical stretch or shift, it is helpful to look for a point on the graph that is relatively clear. In this graph, it appears that $g(2) = 2$. With the basic cubic function at the same input, $f(2) = 2^3 = 8$.

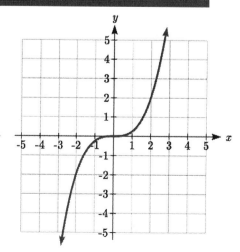

Based on that, it appears that the outputs of g are ¼ the outputs of the function f, since

$$g(2) = \frac{1}{4} f(2).$$

From this we can fairly safely conclude that:

$$g(x) = \frac{1}{4} f(x)$$

We can write a formula for g by using the definition of the function f

$$g(x) = \frac{1}{4} f(x) = \frac{1}{4} x^3$$

Now we consider changes to the inside of a function.

Example 16

Returning to the fruit fly population we looked at earlier, suppose the scientist is now comparing it to a population that progresses through its lifespan twice as fast as the original population. In other words, this new population, R, will progress in 1 hour the same amount the original population did in 2 hours, and in 2 hours, will progress as much as the original population did in 4 hours. Sketch a graph of this population.

Symbolically, we could write
$R(1) = P(2)$
$R(2) = P(4)$, and in general,
$R(t) = P(2t)$

Graphing this,

Original population, $P(t)$

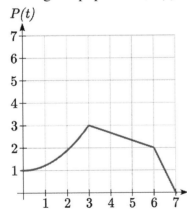

Transformed, $R(t)$

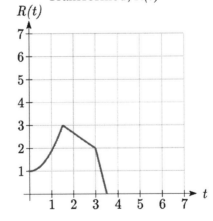

Note the effect on the graph is a horizontal compression, where all input values are half their original distance from the vertical axis.

Horizontal Stretch/Compression

Given a function $f(x)$, if we define a new function $g(x)$ as

$g(x) = f(kx)$, where k is a constant

then $g(x)$ is a **horizontal stretch or compression** of the function $f(x)$.

If $k > 1$, then the graph will be compressed by $1/k$

If $0 < k < 1$, then the graph will be stretched by $1/k$

If $k < 0$, then there will be combination of a horizontal stretch or compression with a horizontal reflection.

Example 17

A function $f(x)$ is given as a table below. Create a table for the function $g(x) = f\left(\dfrac{1}{2}x\right)$

x	2	4	6	8
$f(x)$	1	3	7	11

The formula $g(x) = f\left(\dfrac{1}{2}x\right)$ tells us that the output values for g are the same as the output values for the function f at an input half the size. Notice that we don't have enough information to determine $g(2)$ since $g(2) = f\left(\dfrac{1}{2}\cdot 2\right) = f(1)$, and we do not have a value for $f(1)$ in our table. Our input values to g will need to be twice as large to get inputs for f that we can evaluate. For example, we can determine $g(4)$ since

$g(4) = f\left(\dfrac{1}{2}\cdot 4\right) = f(2) = 1$.

x	4	8	12	16
$g(x)$	1	3	7	11

Since each input value has been doubled, the result is that the function $g(x)$ has been stretched horizontally by 2.

Example 18

Two graphs are shown below. Relate the function $g(x)$ to $f(x)$.

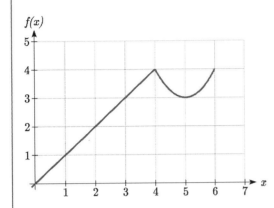
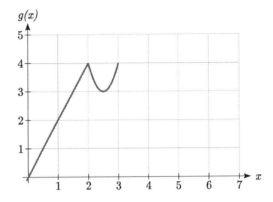

The graph of $g(x)$ looks like the graph of $f(x)$ horizontally compressed. Since $f(x)$ ends at (6,4) and $g(x)$ ends at (2,4) we can see that the x values have been compressed by 1/3, because 6(1/3) = 2. We might also notice that $g(2) = f(6)$, and $g(1) = f(3)$. Either way, we can describe this relationship as $g(x) = f(3x)$. This is a horizontal compression by 1/3.

Notice that the coefficient needed for a horizontal stretch or compression is the *reciprocal* of the stretch or compression. To stretch the graph horizontally by 4, we need a coefficient of 1/4 in our function: $f\left(\dfrac{1}{4}x\right)$. This means the input values must be four times larger to produce the same result, requiring the input to be larger, causing the horizontal stretching.

Try it Now
4. Write a formula for the toolkit square root function horizontally stretched by three.

It is useful to note that for most toolkit functions, a horizontal stretch or vertical stretch can be represented in other ways. For example, a horizontal compression of the function $f(x) = x^2$ by ½ would result in a new function $g(x) = (2x)^2$, but this can also be written as $g(x) = 4x^2$, a vertical stretch of $f(x)$ by 4. When writing a formula for a transformed toolkit, we only need to find one transformation that would produce the graph.

Combining Transformations

When combining transformations, it is very important to consider the order of the transformations. For example, vertically shifting by 3 and then vertically stretching by 2 does not create the same graph as vertically stretching by 2 then vertically shifting by 3.

When we see an expression like $2f(x)+3$, which transformation should we start with? The answer here follows nicely from order of operations, for outside transformations. Given the output value of $f(x)$, we first multiply by 2, causing the vertical stretch, then add 3, causing the vertical shift. (Multiplication before Addition)

Combining Vertical Transformations

When combining vertical transformations written in the form $af(x)+k$,

first vertically stretch by a, then vertically shift by k.

Horizontal transformations are a little trickier to think about. When we write $g(x)=f(2x+3)$ for example, we have to think about how the inputs to the g function relate to the inputs to the f function. Suppose we know $f(7)=12$. What input to g would produce that output? In other words, what value of x will allow $g(x)=f(2x+3)=f(12)$? We would need $2x+3=12$. To solve for x, we would first subtract 3, resulting in horizontal shift, then divide by 2, causing a horizontal compression.

Combining Horizontal Transformations

When **combining horizontal transformations** written in the form $f(bx+p)$,

first horizontally shift by p, then horizontally stretch by $1/b$.

This format ends up being very difficult to work with, since it is usually much easier to horizontally stretch a graph before shifting. We can work around this by factoring inside the function.

$$f(bx+p)=f\left(b\left(x+\frac{p}{b}\right)\right)$$

Factoring in this way allows us to horizontally stretch first, then shift horizontally.

Combining Horizontal Transformations (Factored Form)

When **combining horizontal transformations** written in the form $f(b(x+h))$,

first horizontally stretch by $1/b$, then horizontally shift by h.

Independence of Horizontal and Vertical Transformations

Horizontal and vertical transformations are independent. It does not matter whether horizontal or vertical transformations are done first.

Example 19

Given the table of values for the function $f(x)$ below, create a table of values for the function $g(x) = 2f(3x) + 1$

x	6	12	18	24
$f(x)$	10	14	15	17

There are 3 steps to this transformation and we will work from the inside out. Starting with the horizontal transformations, $f(3x)$ is a horizontal compression by 1/3, which means we multiply each x value by 1/3.

x	2	4	6	8
$f(3x)$	10	14	15	17

Looking now to the vertical transformations, we start with the vertical stretch, which will multiply the output values by 2. We apply this to the previous transformation.

x	2	4	6	8
$2f(3x)$	20	28	30	34

Finally, we can apply the vertical shift, which will add 1 to all the output values.

x	2	4	6	8
$g(x) = 2f(3x) + 1$	21	29	31	35

Example 20

Using the graph of $f(x)$ below, sketch a graph of $k(x) = f\left(\dfrac{1}{2}x + 1\right) - 3$

To make things simpler, we'll start by factoring out the inside of the function

$$f\left(\frac{1}{2}x + 1\right) - 3 = f\left(\frac{1}{2}(x+2)\right) - 3$$

By factoring the inside, we can first horizontally stretch by 2, as indicated by the ½ on the inside of the function. Remember twice the size of 0 is still 0, so the point (0,2) remains at (0,2) while the point (2,0) will stretch to (4,0).

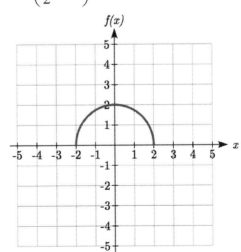

Next, we horizontally shift left by 2 units, as indicated by the *x*+2.

Last, we vertically shift down by 3 to complete our sketch, as indicated by the -3 on the outside of the function.

Horizontal stretch by 2   Horizontal shift left by 2   Vertical shift down 3

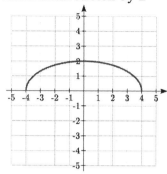   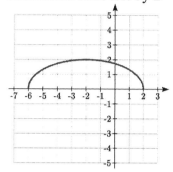   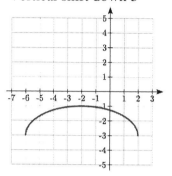

Example 21

Write an equation for the transformed graph of the quadratic function shown.

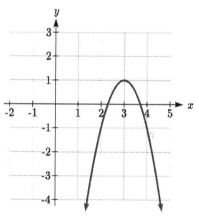

Since this is a quadratic function, first consider what the basic quadratic tool kit function looks like and how this has changed. Observing the graph, we notice several transformations:

The original tool kit function has been flipped over the *x* axis, some kind of stretch or compression has occurred, and we can see a shift to the right 3 units and a shift up 1 unit.

In total there are four operations:
Vertical reflection, requiring a negative sign outside the function
Vertical Stretch *or* Horizontal Compression[*]
Horizontal Shift Right 3 units, which tells us to put *x*-3 on the inside of the function
Vertical Shift up 1 unit, telling us to add 1 on the outside of the function

[*] It is unclear from the graph whether it is showing a vertical stretch or a horizontal compression. For the quadratic, it turns out we could represent it either way, so we'll use a vertical stretch. You may be able to determine the vertical stretch by observation.

By observation, the basic tool kit function has a vertex at (0, 0) and symmetrical points at (1, 1) and (-1, 1). These points are one unit up and one unit over from the vertex. The new points on the transformed graph are one unit away horizontally but 2 units away vertically. They have been stretched vertically by two.

Not everyone can see this by simply looking at the graph. If you can, great, but if not, we can solve for it. First, we will write the equation for this graph, with an unknown vertical stretch.

$f(x) = x^2$ The original function

$-f(x) = -x^2$ Vertically reflected

$-af(x) = -ax^2$ Vertically stretched

$-af(x-3) = -a(x-3)^2$ Shifted right 3

$-af(x-3)+1 = -a(x-3)^2 +1$ Shifted up 1

We now know our graph is going to have an equation of the form $g(x) = -a(x-3)^2 +1$. To find the vertical stretch, we can identify any point on the graph (other than the highest point), such as the point (2,-1), which tells us $g(2) = -1$. Using our general formula, and substituting 2 for x, and -1 for $g(x)$

$-1 = -a(2-3)^2 +1$

$-1 = -a +1$

$-2 = -a$

$2 = a$

This tells us that to produce the graph we need a vertical stretch by two. The function that produces this graph is therefore $g(x) = -2(x-3)^2 +1$.

Try it Now

5. Consider the linear function $g(x) = -2x +1$. Describe its transformation in words using the identity tool kit function $f(x) = x$ as a reference.

Example 22

On what interval(s) is the function $g(x) = \dfrac{-2}{(x-1)^2} +3$ increasing and decreasing?

This is a transformation of the toolkit reciprocal squared function, $f(x) = \dfrac{1}{x^2}$:

$-2f(x) = \dfrac{-2}{x^2}$ A vertical flip and vertical stretch by 2

$-2f(x-1) = \dfrac{-2}{(x-1)^2}$ A shift right by 1

$-2f(x-1)+3 = \dfrac{-2}{(x-1)^2} +3$ A shift up by 3

The basic reciprocal squared function is increasing on $(-\infty,0)$ and decreasing on $(0,\infty)$. Because of the vertical flip, the *g(x)* function will be decreasing on the left and increasing on the right. The horizontal shift right by 1 will also shift these intervals to the right one. From this, we can determine *g(x)* will be increasing on $(1,\infty)$ and decreasing on $(-\infty,1)$. We also could graph the transformation to help us determine these intervals.

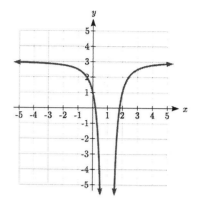

Try it Now

6. On what interval(s) is the function $h(t) = (t-3)^3 + 2$ concave up and down?

Important Topics of This Section
Transformations
Vertical Shift (up & down)
Horizontal Shifts (left & right)
Reflections over the vertical & horizontal axis
Even & Odd functions
Vertical Stretches & Compressions
Horizontal Stretches & Compressions
Combinations of Transformation

Try it Now Answers

1. $b(t) = h(t) + 10 = -4.9t^2 + 30t + 10$

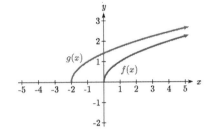

2. a. Horizontal shift
   b. The function is shifted to the LEFT by 2 units.
   c. Shown to the right

3. Shown to the right
   Notice: $g(x) = f(-x)$ looks the same as $f(x)$

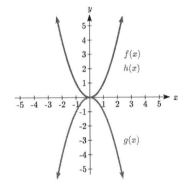

4. $g(x) = f\left(\dfrac{1}{3}x\right)$ so using the square root function we get

   $g(x) = \sqrt{\dfrac{1}{3}x}$

5. The identity tool kit function $f(x) = x$ has been transformed in 3 steps
   a. Vertically stretched by 2.
   b. Vertically reflected over the x axis.
   c. Vertically shifted up by 1 unit.

6. $h(t)$ is concave down on $(-\infty, 3)$ and concave up on $(3, \infty)$

Section 1.5 Exercises

Describe how each function is a transformation of the original function $f(x)$

1. $f(x-49)$ 2. $f(x+43)$

3. $f(x+3)$ 4. $f(x-4)$

5. $f(x)+5$ 6. $f(x)+8$

7. $f(x)-2$ 8. $f(x)-7$

9. $f(x-2)+3$ 10. $f(x+4)-1$

11. Write a formula for $f(x)=\sqrt{x}$ shifted up 1 unit and left 2 units.

12. Write a formula for $f(x)=|x|$ shifted down 3 units and right 1 unit.

13. Write a formula for $f(x)=\dfrac{1}{x}$ shifted down 4 units and right 3 units.

14. Write a formula for $f(x)=\dfrac{1}{x^2}$ shifted up 2 units and left 4 units.

15. Tables of values for $f(x)$, $g(x)$, and $h(x)$ are given below. Write $g(x)$ and $h(x)$ as transformations of $f(x)$.

x	$f(x)$
-2	-2
-1	-1
0	-3
1	1
2	2

x	$g(x)$
-1	-2
0	-1
1	-3
2	1
3	2

x	$h(x)$
-2	-1
-1	0
0	-2
1	2
2	3

16. Tables of values for $f(x)$, $g(x)$, and $h(x)$ are given below. Write $g(x)$ and $h(x)$ as transformations of $f(x)$.

x	$f(x)$
-2	-1
-1	-3
0	4
1	2
2	1

x	$g(x)$
-3	-1
-2	-3
-1	4
0	2
1	1

x	$h(x)$
-2	-2
-1	-4
0	3
1	1
2	0

The graph of $f(x) = 2^x$ is shown. Sketch a graph of each transformation of $f(x)$.

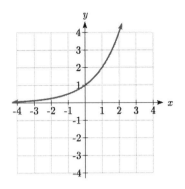

17. $g(x) = 2^x + 1$

18. $h(x) = 2^x - 3$

19. $w(x) = 2^{x-1}$

20. $q(x) = 2^{x+3}$

Sketch a graph of each function as a transformation of a toolkit function.

21. $f(t) = (t+1)^2 - 3$

22. $h(x) = |x-1| + 4$

23. $k(x) = (x-2)^3 - 1$

24. $m(t) = 3 + \sqrt{t+2}$

Write an equation for each function graphed below.

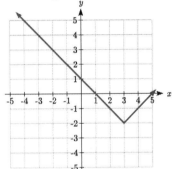

25.

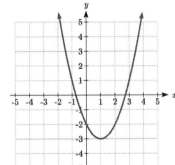

26.

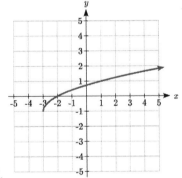

27.

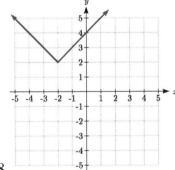

28.

Find a formula for each of the transformations of the square root whose graphs are given below.

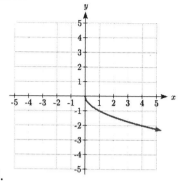

29.

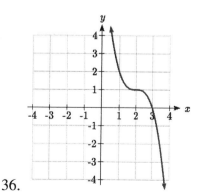

30.

The graph of $f(x) = 2^x$ is shown. Sketch a graph of each transformation of $f(x)$

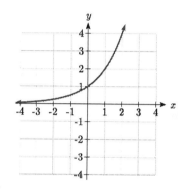

31. $g(x) = -2^x + 1$
32. $h(x) = 2^{-x}$

33. Starting with the graph of $f(x) = 6^x$ write the equation of
    the graph that results from
    a. reflecting $f(x)$ about the x-axis and the y-axis
    b. reflecting $f(x)$ about the x-axis, shifting left 2 units, and down 3 units

34. Starting with the graph of $f(x) = 4^x$ write the equation of the graph that results from
    a. reflecting $f(x)$ about the x-axis
    b. reflecting $f(x)$ about the y-axis, shifting right 4 units, and up 2 units

Write an equation for each function graphed below.

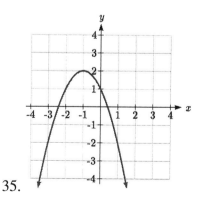

35.

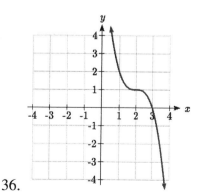

36.

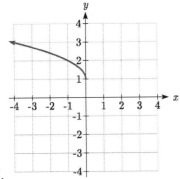

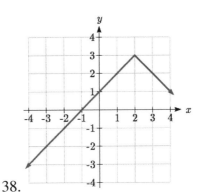

37.

38.

39. For each equation below, determine if the function is Odd, Even, or Neither.

    a. $f(x) = 3x^4$

    b. $g(x) = \sqrt{x}$

    c. $h(x) = \dfrac{1}{x} + 3x$

40. For each equation below, determine if the function is Odd, Even, or Neither.

    a. $f(x) = (x-2)^2$

    b. $g(x) = 2x^4$

    c. $h(x) = 2x - x^3$

Describe how each function is a transformation of the original function $f(x)$.

41. $-f(x)$ 42. $f(-x)$

43. $4f(x)$ 44. $6f(x)$

45. $f(5x)$ 46. $f(2x)$

47. $f\left(\dfrac{1}{3}x\right)$ 48. $f\left(\dfrac{1}{5}x\right)$

49. $3f(-x)$ 50. $-f(3x)$

Write a formula for the function that results when the given toolkit function is transformed as described.

51. $f(x) = |x|$ reflected over the y axis and horizontally compressed by a factor of $\dfrac{1}{4}$.

52. $f(x) = \sqrt{x}$ reflected over the x axis and horizontally stretched by a factor of 2.

53. $f(x) = \dfrac{1}{x^2}$ vertically compressed by a factor of $\dfrac{1}{3}$, then shifted to the left 2 units and down 3 units.

54. $f(x) = \dfrac{1}{x}$ vertically stretched by a factor of 8, then shifted to the right 4 units and up 2 units.

55. $f(x) = x^2$ horizontally compressed by a factor of $\dfrac{1}{2}$, then shifted to the right 5 units and up 1 unit.

56. $f(x) = x^2$ horizontally stretched by a factor of 3, then shifted to the left 4 units and down 3 units.

Describe how each formula is a transformation of a toolkit function. Then sketch a graph of the transformation.

57. $f(x) = 4(x+1)^2 - 5$

58. $g(x) = 5(x+3)^2 - 2$

59. $h(x) = -2|x-4| + 3$

60. $k(x) = -3\sqrt{x} - 1$

61. $m(x) = \dfrac{1}{2}x^3$

62. $n(x) = \dfrac{1}{3}|x-2|$

63. $p(x) = \left(\dfrac{1}{3}x\right)^2 - 3$

64. $q(x) = \left(\dfrac{1}{4}x\right)^3 + 1$

65. $a(x) = \sqrt{-x+4}$

66. $b(x) = \sqrt[3]{-x-6}$

Determine the interval(s) on which the function is increasing and decreasing.

67. $f(x) = 4(x+1)^2 - 5$

68. $g(x) = 5(x+3)^2 - 2$

69. $a(x) = \sqrt{-x+4}$

70. $k(x) = -3\sqrt{x} - 1$

Determine the interval(s) on which the function is concave up and concave down.

71. $m(x) = -2(x+3)^3 + 1$

72. $b(x) = \sqrt[3]{-x-6}$

73. $p(x) = \left(\dfrac{1}{3}x\right)^2 - 3$

74. $k(x) = -3\sqrt{x} - 1$

The function $f(x)$ is graphed here. Write an equation for each graph below as a transformation of $f(x)$.

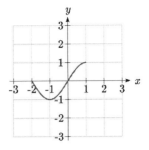

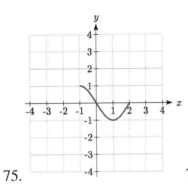

75.

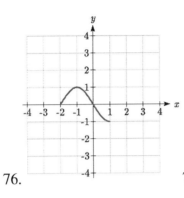

76.

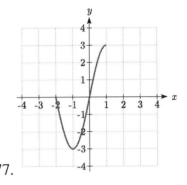

77.

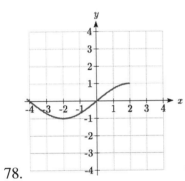

78.

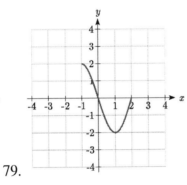

79.

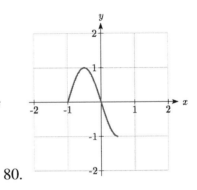

80.

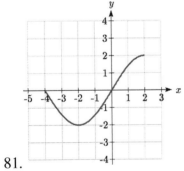

81.

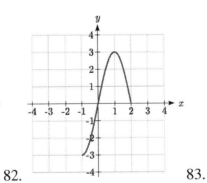

82.

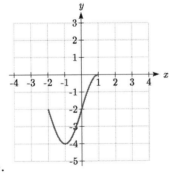

83.

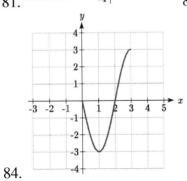

84.

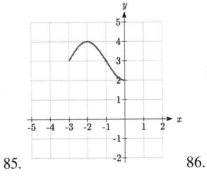

85.

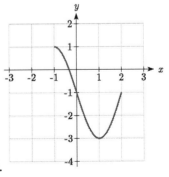

86.

Write an equation for each transformed toolkit function graphed below.

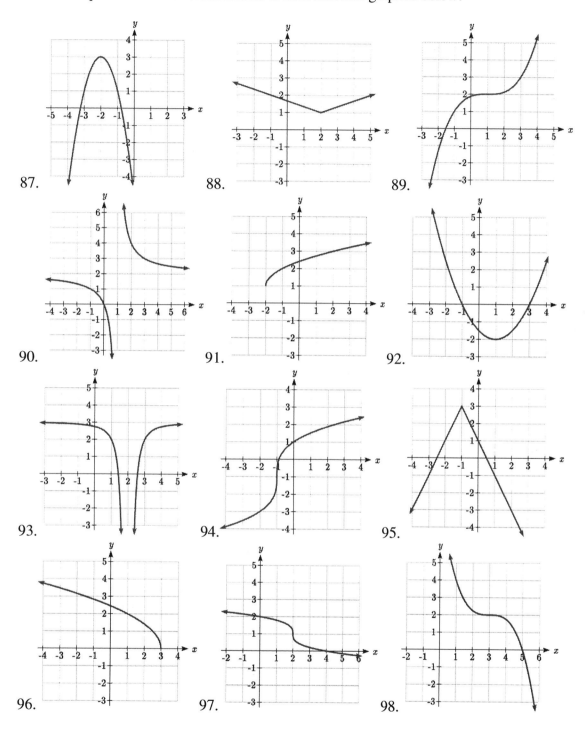

87.

88.

89.

90.

91.

92.

93.

94.

95.

96.

97.

98.

Write a formula for the piecewise function graphed below.

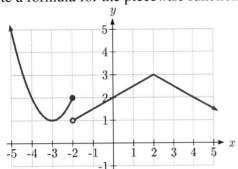

99.

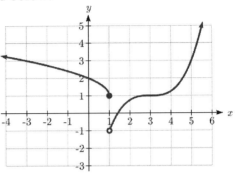

100.

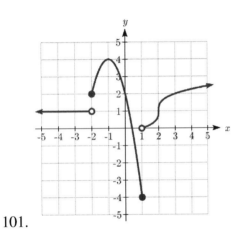

101.

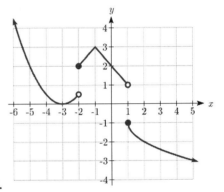

102.

103. Suppose you have a function $y = f(x)$ such that the domain of $f(x)$ is $1 \leq x \leq 6$ and the range of $f(x)$ is $-3 \leq y \leq 5$. [UW]

    a. What is the domain of $f(2(x-3))$?

    b. What is the range of $f(2(x-3))$?

    c. What is the domain of $2f(x)-3$?

    d. What is the range of $2f(x)-3$?

    e. Can you find constants B and C so that the domain of $f(B(x-C))$ is $8 \leq x \leq 9$?

    f. Can you find constants A and D so that the range of $Af(x)+D$ is $0 \leq y \leq 1$?

Section 1.6 Inverse Functions

A fashion designer is travelling to Milan for a fashion show. He asks his assistant, Betty, what 75 degrees Fahrenheit is in Celsius, and after a quick search on Google, she finds the formula $C = \frac{5}{9}(F - 32)$. Using this formula, she calculates $\frac{5}{9}(75 - 32) \approx 24$ degrees Celsius. The next day, the designer sends his assistant the week's weather forecast for Milan, and asks her to convert the temperatures to Fahrenheit.

Mon	Tue	Wed	Thu
26°C | 19°C	29°C | 19°C	30°C | 20°C	26°C | 18°C

At first, Betty might consider using the formula she has already found to do the conversions. After all, she knows her algebra, and can easily solve the equation for F after substituting a value for C. For example, to convert 26 degrees Celsius, she could write:

$$26 = \frac{5}{9}(F - 32)$$

$$26 \cdot \frac{9}{5} = F - 32$$

$$F = 26 \cdot \frac{9}{5} + 32 \approx 79$$

After considering this option for a moment, she realizes that solving the equation for each of the temperatures would get awfully tedious, and realizes that since evaluation is easier than solving, it would be much more convenient to have a different formula, one which takes the Celsius temperature and outputs the Fahrenheit temperature. This is the idea of an inverse function, where the input becomes the output and the output becomes the input.

Inverse Function

If $f(a) = b$, then a function $g(x)$ is an **inverse** of f if $g(b) = a$.

The inverse of $f(x)$ is typically notated $f^{-1}(x)$, which is read "f inverse of x", so equivalently, if $f(a) = b$ then $f^{-1}(b) = a$.

Important: The raised -1 used in the notation for inverse functions is simply a notation, and does not designate an exponent or power of -1.

Example 1

If for a particular function, $f(2) = 4$, what do we know about the inverse?

The inverse function reverses which quantity is input and which quantity is output, so if $f(2) = 4$, then $f^{-1}(4) = 2$.

Alternatively, if you want to re-name the inverse function $g(x)$, then $g(4) = 2$

Try it Now

1. Given that $h^{-1}(6) = 2$, what do we know about the original function $h(x)$?

Notice that original function and the inverse function *undo* each other. If $f(a) = b$, then $f^{-1}(b) = a$, returning us to the original input. More simply put, if you compose these functions together you get the original input as your answer.

$$f^{-1}(f(a)) = a \quad \text{and} \quad f(f^{-1}(b)) = b$$

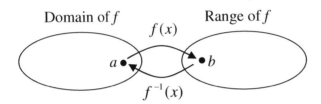

Since the outputs of the function f are the inputs to f^{-1}, the range of f is also the domain of f^{-1}. Likewise, since the inputs to f are the outputs of f^{-1}, the domain of f is the range of f^{-1}.

Basically, like how the input and output values switch, the domain & ranges switch as well. But be careful, because sometimes a function doesn't even have an inverse function, or only has an inverse on a limited domain. For example, the inverse of $f(x) = \sqrt{x}$ is $f^{-1}(x) = x^2$, since a square "undoes" a square root, but it is only the inverse of $f(x)$ on the domain $[0, \infty)$, since that is the range of $f(x) = \sqrt{x}$.

Example 2

The function $f(x) = 2^x$ has domain $(-\infty, \infty)$ and range $(0, \infty)$, what would we expect the domain and range of f^{-1} to be?

We would expect f^{-1} to swap the domain and range of f, so f^{-1} would have domain $(0, \infty)$ and range $(-\infty, \infty)$.

Example 3

A function $f(t)$ is given as a table below, showing distance in miles that a car has traveled in t minutes. Find and interpret $f^{-1}(70)$

t (minutes)	30	50	70	90
$f(t)$ (miles)	20	40	60	70

The inverse function takes an output of f and returns an input for f. So in the expression $f^{-1}(70)$, the 70 is an output value of the original function, representing 70 miles. The inverse will return the corresponding input of the original function f, 90 minutes, so $f^{-1}(70) = 90$. Interpreting this, it means that to drive 70 miles, it took 90 minutes.

Alternatively, recall the definition of the inverse was that if $f(a) = b$ then $f^{-1}(b) = a$. By this definition, if you are given $f^{-1}(70) = a$ then you are looking for a value a so that $f(a) = 70$. In this case, we are looking for a t so that $f(t) = 70$, which is when $t = 90$.

Try it Now

2. Using the table below

t (minutes)	30	50	60	70	90
$f(t)$ (miles)	20	40	50	60	70

Find and interpret the following
   a. $f(60)$
   b. $f^{-1}(60)$

Example 4

A function $g(x)$ is given as a graph below. Find $g(3)$ and $g^{-1}(3)$

To evaluate $g(3)$, we find 3 on the horizontal axis and find the corresponding output value on the vertical axis. The point (3, 1) tells us that $g(3) = 1$

To evaluate $g^{-1}(3)$, recall that by definition $g^{-1}(3)$ means $g(x) = 3$. By looking for the output value 3 on the vertical axis we find the point (5, 3) on the graph, which means $g(5) = 3$, so by definition $g^{-1}(3) = 5$.

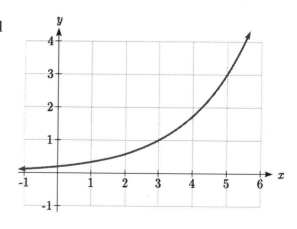

3. Using the graph in Example 4 above

a. find $g^{-1}(1)$

b. estimate $g^{-1}(4)$

Example 5

Returning to our designer's assistant, find a formula for the inverse function that gives Fahrenheit temperature given a Celsius temperature.

A quick Google search would find the inverse function, but alternatively, Betty might look back at how she solved for the Fahrenheit temperature for a specific Celsius value, and repeat the process in general

$$C = \frac{5}{9}(F - 32)$$

$$C \cdot \frac{9}{5} = F - 32$$

$$F = \frac{9}{5}C + 32$$

By solving in general, we have uncovered the inverse function. If

$$C = h(F) = \frac{5}{9}(F - 32)$$

Then

$$F = h^{-1}(C) = \frac{9}{5}C + 32$$

In this case, we introduced a function h to represent the conversion since the input and output variables are descriptive, and writing C^{-1} could get confusing.

It is important to note that not all functions will have an inverse function. Since the inverse $f^{-1}(x)$ takes an output of f and returns an input of f, in order for f^{-1} to itself be a function, then each output of f (input to f^{-1}) must correspond to exactly one input of f (output of f^{-1}) in order for f^{-1} to be a function. You might recall that this is the definition of a one-to-one function.

Properties of Inverses

In order **for a function to have an inverse**, it must be a one-to-one function.

In some cases, it is desirable to have an inverse for a function even though the function is not one-to-one. In those cases, we can often limit the domain of the original function to an interval on which the function *is* one-to-one, then find an inverse only on that interval.

If you have not already done so, go back to the toolkit functions that were not one-to-one and limit or restrict the domain of the original function so that it is one-to-one. If you are not sure how to do this, proceed to Example 6.

Example 6

The quadratic function $h(x) = x^2$ is not one-to-one. Find a domain on which this function is one-to-one, and find the inverse on that domain.

We can limit the domain to $[0, \infty)$ to restrict the graph to a portion that is one-to-one, and find an inverse on this limited domain.

You may have already guessed that since we undo a square with a square root, the inverse of $h(x) = x^2$ on this domain is $h^{-1}(x) = \sqrt{x}$.

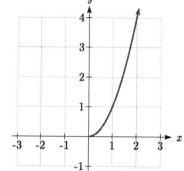

You can also solve for the inverse function algebraically. If $h(x) = x^2$, we can introduce the variable y to represent the output values, allowing us to write $y = x^2$. To find the inverse we solve for the input variable

To solve for x we take the square root of each side. $\sqrt{y} = \sqrt{x^2}$ and get $\sqrt{y} = |x|$, so $x = \pm\sqrt{y}$. We have restricted x to being non-negative, so we'll use the positive square root, $x = \sqrt{y}$ or $h^{-1}(y) = \sqrt{y}$. In cases like this where the variables are not descriptive, it is common to see the inverse function rewritten with the variable x: $h^{-1}(x) = \sqrt{x}$. Rewriting the inverse using the variable x is often required for graphing inverse functions using calculators or computers.

Note that the domain and range of the square root function do correspond with the range and domain of the quadratic function on the limited domain. In fact, if we graph $h(x)$ on the restricted domain and $h^{-1}(x)$ on the same axes, we can notice symmetry: the graph of $h^{-1}(x)$ is the graph of $h(x)$ reflected over the line $y = x$.

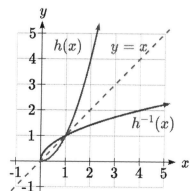

Example 7

Given the graph of *f(x)* shown, sketch a graph of $f^{-1}(x)$.

This is a one-to-one function, so we will be able to sketch an inverse. Note that the graph shown has an apparent domain of $(0,\infty)$ and range of $(-\infty,\infty)$, so the inverse will have a domain of $(-\infty,\infty)$ and range of $(0,\infty)$.

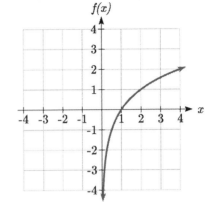

Reflecting this graph of the line $y = x$, the point $(1, 0)$ reflects to $(0, 1)$, and the point $(4, 2)$ reflects to $(2, 4)$. Sketching the inverse on the same axes as the original graph:

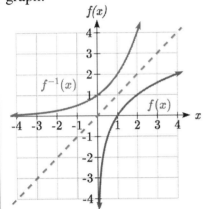

Important Topics of this Section
Definition of an inverse function
Composition of inverse functions yield the original input value
Not every function has an inverse function
To have an inverse a function must be one-to-one
Restricting the domain of functions that are not one-to-one.

Try it Now Answers

1. $g(2) = 6$

2.a. $f(60) = 50$. In 60 minutes, 50 miles are traveled.

   b. $f^{-1}(60) = 70$. To travel 60 miles, it will take 70 minutes.

3. a. $g^{-1}(1) = 3$

   b. $g^{-1}(4) = 5.5$ (this is an approximation – answers may vary slightly)

Section 1.6 Exercises

Assume that the function f is a one-to-one function.

1. If $f(6) = 7$, find $f^{-1}(7)$

2. If $f(3) = 2$, find $f^{-1}(2)$

3. If $f^{-1}(-4) = -8$, find $f(-8)$

4. If $f^{-1}(-2) = -1$, find $f(-1)$

5. If $f(5) = 2$, find $\left(f(5)\right)^{-1}$

6. If $f(1) = 4$, find $\left(f(1)\right)^{-1}$

7. Using the graph of $f(x)$ shown

    a. Find $f(0)$

    b. Solve $f(x) = 0$

    c. Find $f^{-1}(0)$

    d. Solve $f^{-1}(x) = 0$

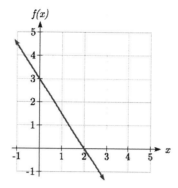

8. Using the graph shown

    a. Find $g(1)$

    b. Solve $g(x) = 1$

    c. Find $g^{-1}(1)$

    d. Solve $g^{-1}(x) = 1$

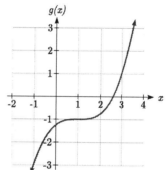

9. Use the table below to find the indicated quantities.

x	0	1	2	3	4	5	6	7	8	9
$f(x)$	8	0	7	4	2	6	5	3	9	1

    a. Find $f(1)$

    b. Solve $f(x) = 3$

    c. Find $f^{-1}(0)$

    d. Solve $f^{-1}(x) = 7$

10. Use the table below to fill in the missing values.

t	0	1	2	3	4	5	6	7	8
h(t)	6	0	1	7	2	3	5	4	9

  a. Find $h(6)$

  b. Solve $h(t) = 0$

  c. Find $h^{-1}(5)$

  d. Solve $h^{-1}(t) = 1$

For each table below, create a table for $f^{-1}(x)$.

11.

x	3	6	9	13	14
f(x)	1	4	7	12	16

12.

x	3	5	7	13	15
f(x)	2	6	9	11	16

For each function below, find $f^{-1}(x)$

13. $f(x) = x + 3$

14. $f(x) = x + 5$

15. $f(x) = 2 - x$

16. $f(x) = 3 - x$

17. $f(x) = 11x + 7$

18. $f(x) = 9 + 10x$

For each function, find a domain on which f is one-to-one and non-decreasing, then find the inverse of f restricted to that domain.

19. $f(x) = (x + 7)^2$

20. $f(x) = (x - 6)^2$

21. $f(x) = x^2 - 5$

22. $f(x) = x^2 + 1$

23. If $f(x) = x^3 - 5$ and $g(x) = \sqrt[3]{x + 5}$, find

  a. $f(g(x))$

  b. $g(f(x))$

  c. What does this tell us about the relationship between $f(x)$ and $g(x)$?

24. If $f(x) = \dfrac{x}{2 + x}$ and $g(x) = \dfrac{2x}{1 - x}$, find

  a. $f(g(x))$

  b. $g(f(x))$

  c. What does this tell us about the relationship between $f(x)$ and $g(x)$?

Chapter 2:
Linear Functions

Chapter one was a window that gave us a peek into the entire course. Our goal was to understand the basic structure of functions and function notation, the toolkit functions, domain and range, how to recognize and understand composition and transformations of functions and how to understand and utilize inverse functions. With these basic components in hand we will further research the specific details and intricacies of each type of function in our toolkit and use them to model the world around us.

Mathematical Modeling

As we approach day to day life we often need to quantify the things around us, giving structure and numeric value to various situations. This ability to add structure enables us to make choices based on patterns we see that are weighted and systematic. With this structure in place we can model and even predict behavior to make decisions. Adding a numerical structure to a real world situation is called **Mathematical Modeling.**

When modeling real world scenarios, there are some common growth patterns that are regularly observed. We will devote this chapter and the rest of the book to the study of the functions used to model these growth patterns.

Section 2.1 Linear Functions

As you hop into a taxicab in Las Vegas, the meter will immediately read $3.50; this is the "drop" charge made when the taximeter is activated. After that initial fee, the taximeter will add $2.76 for each mile the taxi drives[1]. In this scenario, the total taxi fare depends upon the number of miles ridden in the taxi, and we can ask whether it is possible to model this type of scenario with a function. Using descriptive variables, we choose m for miles and C for Cost in dollars as a function of miles: $C(m)$.

---

[1] Nevada Taxicab Authority, retrieved May 8, 2017. There is also a waiting fee assessed when the taxi is waiting at red lights, but we'll ignore that in this discussion.
This chapter is part of *Precalculus: An Investigation of Functions* © Lippman & Rasmussen 2017.
This material is licensed under a Creative Commons CC-BY-SA license.

We know for certain that $C(0) = 3.50$, since the \$3.50 drop charge is assessed regardless of how many miles are driven. Since \$2.67 is added for each mile driven, then $C(1) = 3.50 + 2.67 = 6.17$.

If we then drove a second mile, another \$2.67 would be added to the cost:
$C(2) = 3.50 + 2.67 + 2.67 = 3.50 + 2.67(2) = 8.84$

If we drove a third mile, another \$2.67 would be added to the cost:
$C(3) = 3.50 + 2.67 + 2.67 + 2.67 = 3.50 + 2.67(3) = 11.51$

From this we might observe the pattern, and conclude that if m miles are driven, $C(m) = 3.50 + 2.67m$ because we start with a \$3.50 drop fee and then for each mile increase we add \$2.67.

It is good to verify that the units make sense in this equation. The \$3.50 drop charge is measured in dollars; the \$2.67 charge is measured in dollars per mile.

$$C(m) = 3.50 dollars + \left(2.67 \frac{dollars}{mile} \right)(m \, miles)$$

When dollars per mile are multiplied by a number of miles, the result is a number of dollars, matching the units on the 3.50, and matching the desired units for the C function.

Notice this equation $C(m) = 3.50 + 2.67m$ consisted of two quantities. The first is the fixed \$3.50 charge which does not change based on the value of the input. The second is the \$2.67 dollars per mile value, which is a **rate of change**. In the equation, this rate of change is multiplied by the input value.

Looking at this same problem in table format we can also see the cost changes by \$2.67 for every 1 mile increase.

m	0	1	2	3
$C(m)$	3.50	6.17	8.84	11.51

It is important here to note that in this equation, the **rate of change is constant**; over any interval, the rate of change is the same.

Graphing this equation, $C(m) = 3.50 + 2.67m$ we see the shape is a line, which is how these functions get their name: **linear functions**.

When the number of miles is zero the cost is \$3.50, giving the point (0, 3.50) on the graph. This is the vertical or $C(m)$ intercept. The graph is increasing in a straight line from left to right because for each mile the cost goes up by \$2.67; this rate remains consistent.

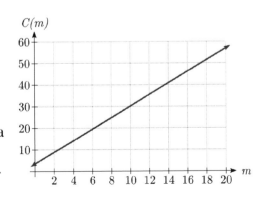

In this example, you have seen the taxicab cost modeled in words, an equation, a table and in graphical form. Whenever possible, ensure that you can link these four representations together to continually build your skills. It is important to note that you will not always be able to find all 4 representations for a problem and so being able to work with all 4 forms is very important.

Linear Function

A **linear function** is a function whose graph produces a line. Linear functions can always be written in the form

$$f(x) = b + mx \qquad \text{or} \qquad f(x) = mx + b \text{; they're equivalent}$$

where

b is the initial or starting value of the function (when input, $x = 0$), and

m is the constant rate of change of the function

Many people like to write linear functions in the form $f(x) = b + mx$ because it corresponds to the way we tend to speak: "The output starts at b and increases at a rate of m."

For this reason alone we will use the $f(x) = b + mx$ form for many of the examples, but remember they are equivalent and can be written correctly both ways.

Slope and Increasing/Decreasing

m is the constant rate of change of the function (also called **slope**). The slope determines if the function is an increasing function or a decreasing function.

$f(x) = b + mx$ is an **increasing** function if $m > 0$

$f(x) = b + mx$ is a **decreasing** function if $m < 0$

If $m = 0$, the rate of change zero, and the function $f(x) = b + 0x = b$ is just a horizontal line passing through the point (0, b), neither increasing nor decreasing.

Example 1

Marcus currently owns 200 songs in his iTunes collection. Every month, he adds 15 new songs. Write a formula for the number of songs, N, in his iTunes collection as a function of the number of months, m. How many songs will he own in a year?

The initial value for this function is 200, since he currently owns 200 songs, so $N(0) = 200$. The number of songs increases by 15 songs per month, so the rate of change is 15 songs per month. With this information, we can write the formula:
$N(m) = 200 + 15m$.

$N(m)$ is an increasing linear function. With this formula we can predict how many songs he will have in 1 year (12 months):

$N(12) = 200 + 15(12) = 200 + 180 = 380$. Marcus will have 380 songs in 12 months.

Try it Now

1. If you earn $30,000 per year and you spend $29,000 per year write an equation for the amount of money you save after *y* years, if you start with nothing.

   "The most important thing, spend less than you earn![2]"

Calculating Rate of Change

Given two values for the input, x_1 and x_2, and two corresponding values for the output, y_1 and y_2, or a set of points, $(x_1,\ y_1)$ and $(x_2,\ y_2)$, if we wish to find a linear function that contains both points we can calculate the rate of change, m:

$$m = \frac{\text{change in output}}{\text{change in input}} = \frac{\Delta y}{\Delta x} = \frac{y_2 - y_1}{x_2 - x_1}$$

Rate of change of a linear function is also called the **slope** of the line.

Note in function notation, $y_1 = f(x_1)$ and $y_2 = f(x_2)$, so we could equivalently write

$$m = \frac{f(x_2) - f(x_1)}{x_2 - x_1}$$

Example 2

The population of a city increased from 23,400 to 27,800 between 2002 and 2006. Find the rate of change of the population during this time span.

The rate of change will relate the change in population to the change in time. The population increased by $27800 - 23400 = 4400$ people over the 4 year time interval. To find the rate of change, the number of people per year the population changed by:

$$\frac{4400\,people}{4\,years} = 1100\frac{people}{year} = 1100 \text{ people per year}$$

Notice that we knew the population was increasing, so we would expect our value for m to be positive. This is a quick way to check to see if your value is reasonable.

---

[2] http://www.thesimpledollar.com/2009/06/19/rule-1-spend-less-than-you-earn/

Example 3

The pressure, P, in pounds per square inch (PSI) on a diver depends upon their depth below the water surface, d, in feet, following the equation $P(d) = 14.696 + 0.434d$. Interpret the components of this function.

The rate of change, or slope, 0.434 would have units $\dfrac{\text{output}}{\text{input}} = \dfrac{\text{pressure}}{\text{depth}} = \dfrac{\text{PSI}}{\text{ft}}$. This tells us the pressure on the diver increases by 0.434 PSI for each foot their depth increases.

The initial value, 14.696, will have the same units as the output, so this tells us that at a depth of 0 feet, the pressure on the diver will be 14.696 PSI.

Example 4

If $f(x)$ is a linear function, $f(3) = -2$, and $f(8) = 1$, find the rate of change.

$f(3) = -2$ tells us that the input 3 corresponds with the output -2, and $f(8) = 1$ tells us that the input 8 corresponds with the output 1. To find the rate of change, we divide the change in output by the change in input:

$m = \dfrac{\text{change in output}}{\text{change in input}} = \dfrac{1 - (-2)}{8 - 3} = \dfrac{3}{5}$. If desired we could also write this as $m = 0.6$

Note that it is not important which pair of values comes first in the subtractions so long as the first output value used corresponds with the first input value used.

Try it Now

2. Given the two points (2, 3) and (0, 4), find the rate of change. Is this function increasing or decreasing?

We can now find the rate of change given two input-output pairs, and can write an equation for a linear function once we have the rate of change and initial value. If we have two input-output pairs and they do not include the initial value of the function, then we will have to solve for it.

Example 5

Write an equation for the linear function graphed to the right.

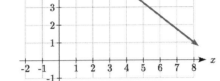

Looking at the graph, we might notice that it passes through the points $(0, 7)$ and $(4, 4)$. From the first value, we know the initial value of the function is $b = 7$, so in this case we will only need to calculate the rate of change:

$$m = \frac{4 - 7}{4 - 0} = \frac{-3}{4}$$

This allows us to write the equation:

$$f(x) = 7 - \frac{3}{4}x$$

Example 6

If $f(x)$ is a linear function, $f(3) = -2$, and $f(8) = 1$, find an equation for the function.

In example 3, we computed the rate of change to be $m = \frac{3}{5}$. In this case, we do not know the initial value $f(0)$, so we will have to solve for it. Using the rate of change, we know the equation will have the form $f(x) = b + \frac{3}{5}x$. Since we know the value of the function when $x = 3$, we can evaluate the function at 3.

$$f(3) = b + \frac{3}{5}(3)$$ Since we know that $f(3) = -2$, we can substitute on the left side

$$-2 = b + \frac{3}{5}(3)$$ This leaves us with an equation we can solve for the initial value

$$b = -2 - \frac{9}{5} = \frac{-19}{5}$$

Combining this with the value for the rate of change, we can now write a formula for this function:

$$f(x) = \frac{-19}{5} + \frac{3}{5}x$$

Example 7

Working as an insurance salesperson, Ilya earns a base salary and a commission on each new policy, so Ilya's weekly income, I, depends on the number of new policies, n, he sells during the week. Last week he sold 3 new policies, and earned $760 for the week. The week before, he sold 5 new policies, and earned $920. Find an equation for $I(n)$, and interpret the meaning of the components of the equation.

The given information gives us two input-output pairs: (3,760) and (5,920). We start by finding the rate of change.

$$m = \frac{920 - 760}{5 - 3} = \frac{160}{2} = 80$$

Keeping track of units can help us interpret this quantity. Income increased by $160 when the number of policies increased by 2, so the rate of change is $80 per policy; Ilya earns a commission of $80 for each policy sold during the week.

We can then solve for the initial value

$I(n) = b + 80n$ then when $n = 3$, $I(3) = 760$, giving

$760 = b + 80(3)$ this allows us to solve for b

$b = 760 - 80(3) = 520$

This value is the starting value for the function. This is Ilya's income when $n = 0$, which means no new policies are sold. We can interpret this as Ilya's base salary for the week, which does not depend upon the number of policies sold.

Writing the final equation:

$I(n) = 520 + 80n$

Our final interpretation is: Ilya's base salary is $520 per week and he earns an additional $80 commission for each policy sold each week.

Flashback

Looking at Example 7:
    Determine the independent and dependent variables.
    What is a reasonable domain and range?
    Is this function one-to-one?

Try it Now

3. The balance in your college payment account, C, is a function of the number of quarters, q, you attend. Interpret the function $C(a) = 20000 - 4000q$ in words. How many quarters of college can you pay for until this account is empty?

Example 8

Given the table below write a linear equation that represents the table values

w, number of weeks	0	2	4	6
$P(w)$, number of rats	1000	1080	1160	1240

We can see from the table that the initial value of rats is 1000 so in the linear format $P(w) = b + mw$, $b = 1000$.

Rather than solving for m, we can notice from the table that the population goes up by 80 for every 2 weeks that pass. This rate is consistent from week 0, to week 2, 4, and 6. The rate of change is 80 rats per 2 weeks. This can be simplified to 40 rats per week and we can write

$P(w) = b + mw$ as $P(w) = 1000 + 40w$

If you didn't notice this from the table you could still solve for the slope using any two points from the table. For example, using (2, 1080) and (6, 1240),

$$m = \frac{1240 - 1080}{6 - 2} = \frac{160}{4} = 40 \text{ rats per week}$$

Important Topics of this Section

Definition of Modeling

Definition of a linear function

Structure of a linear function

Increasing & Decreasing functions

Finding the vertical intercept $(0, b)$

Finding the slope/rate of change, m

Interpreting linear functions

Try it Now Answers

1. $S(y) = 30,000 y - 29,000 y = 1000 y$ $1000 is saved each year.

2. $m = \dfrac{4-3}{0-2} = \dfrac{1}{-2} = -\dfrac{1}{2}$; Decreasing because $m < 0$

3. Your College account starts with $20,000 in it and you withdraw $4,000 each quarter (or your account contains $20,000 and decreases by $4000 each quarter.)
   Solving $C(a) = 0$ gives $a = 5$. You can pay for 5 quarters before the money in this account is gone.

Flashback Answers

n (number of policies sold) is the independent variable

I(n) (weekly income as a function of policies sold) is the dependent variable.

A reasonable domain is $(0, 15)^*$

A reasonable range is $(\$540, \$1740)^*$

*answers may vary given reasoning is stated; 15 is an arbitrary upper limit based on selling 3 policies per day in a 5 day work week and $1740 corresponds with the domain.

Yes this function is one-to-one

Section 2.1 Exercises

1. A town's population has been growing linearly. In 2003, the population was 45,000, and the population has been growing by 1700 people each year. Write an equation, $P(t)$, for the population t years after 2003.

2. A town's population has been growing linearly. In 2005, the population was 69,000, and the population has been growing by 2500 people each year. Write an equation, $P(t)$, for the population t years after 2005.

3. Sonya is currently 10 miles from home, and is walking further away at 2 miles per hour. Write an equation for her distance from home t hours from now.

4. A boat is 100 miles away from the marina, sailing directly towards it at 10 miles per hour. Write an equation for the distance of the boat from the marina after t hours.

5. Timmy goes to the fair with $40. Each ride costs $2. How much money will he have left after riding n rides?

6. At noon, a barista notices she has $20 in her tip jar. If she makes an average of $0.50 from each customer, how much will she have in her tip jar if she serves n more customers during her shift?

Determine if each function is increasing or decreasing

7. $f(x) = 4x + 3$

8. $g(x) = 5x + 6$

9. $a(x) = 5 - 2x$

10. $b(x) = 8 - 3x$

11. $h(x) = -2x + 4$

12. $k(x) = -4x + 1$

13. $j(x) = \frac{1}{2}x - 3$

14. $p(x) = \frac{1}{4}x - 5$

15. $n(x) = -\frac{1}{3}x - 2$

16. $m(x) = -\frac{3}{8}x + 3$

Find the slope of the line that passes through the two given points

17. (2, 4) and (4, 10)

18. (1, 5) and (4, 11)

19. (-1,4) and (5, 2)

20. (-2, 8) and (4, 6)

21. (6,11) and (-4,3)

22. (9,10) and (-6,-12)

Find the slope of the lines graphed

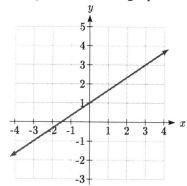

23.

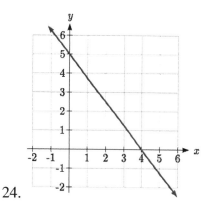

24.

25. Sonya is walking home from a friend's house. After 2 minutes she is 1.4 miles from home. Twelve minutes after leaving, she is 0.9 miles from home. What is her rate?

26. A gym membership with two personal training sessions costs $125, while gym membership with 5 personal training sessions costs $260. What is the rate for personal training sessions?

27. A city's population in the year 1960 was 287,500. In 1989 the population was 275,900. Compute the slope of the population growth (or decline) and make a statement about the population rate of change in people per year.

28. A city's population in the year 1958 was 2,113,000. In 1991 the population was 2,099,800. Compute the slope of the population growth (or decline) and make a statement about the population rate of change in people per year.

29. A phone company charges for service according to the formula: $C(n) = 24 + 0.1n$, where n is the number of minutes talked, and $C(n)$ is the monthly charge, in dollars. Find and interpret the rate of change and initial value.

30. A phone company charges for service according to the formula: $C(n) = 26 + 0.04n$, where n is the number of minutes talked, and $C(n)$ is the monthly charge, in dollars. Find and interpret the rate of change and initial value.

31. Terry is skiing down a steep hill. Terry's elevation, $E(t)$, in feet after t seconds is given by $E(t) = 3000 - 70t$. Write a complete sentence describing Terry's starting elevation and how it is changing over time.

32. Maria is climbing a mountain. Maria's elevation, $E(t)$, in feet after t minutes is given by $E(t)=1200+40t$. Write a complete sentence describing Maria's starting elevation and how it is changing over time.

Given each set of information, find a linear equation satisfying the conditions, if possible

33. $f(-5)=-4$, and $f(5)=2$

34. $f(-1)=4$, and $f(5)=1$

35. Passes through (2, 4) and (4, 10)

36. Passes through (1, 5) and (4, 11)

37. Passes through (-1,4) and (5, 2)

38. Passes through (-2, 8) and (4, 6)

39. x intercept at (-2, 0) and y intercept at (0, -3)

40. x intercept at (-5, 0) and y intercept at (0, 4)

Find an equation for the function graphed

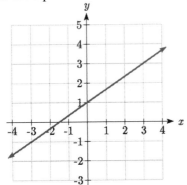

41.

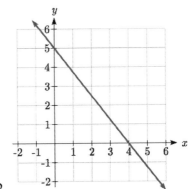

42.

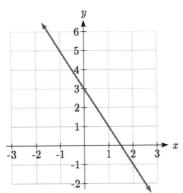

43.

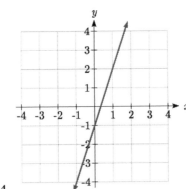

44.

45. A clothing business finds there is a linear relationship between the number of shirts, n, it can sell and the price, p, it can charge per shirt. In particular, historical data shows that 1000 shirts can be sold at a price of \$30, while 3000 shirts can be sold at a price of \$22. Find a linear equation in the form $p=mn+b$ that gives the price p they can charge for n shirts.

46. A farmer finds there is a linear relationship between the number of bean stalks, n, she plants and the yield, y, each plant produces. When she plants 30 stalks, each plant yields 30 oz of beans. When she plants 34 stalks, each plant produces 28 oz of beans. Find a linear relationships in the form $y = mn + b$ that gives the yield when n stalks are planted.

47. Which of the following tables could represent a linear function? For each that could be linear, find a linear equation models the data.

x	$g(x)$
0	5
5	-10
10	-25
15	-40

x	$h(x)$
0	5
5	30
10	105
15	230

x	$f(x)$
0	-5
5	20
10	45
15	70

x	$k(x)$
5	13
10	28
20	58
25	73

48. Which of the following tables could represent a linear function? For each that could be linear, find a linear equation models the data.

x	$g(x)$
0	6
2	-19
4	-44
6	-69

x	$h(x)$
2	13
4	23
8	43
10	53

x	$f(x)$
2	-4
4	16
6	36
8	56

x	$k(x)$
0	6
2	31
6	106
8	231

49. While speaking on the phone to a friend in Oslo, Norway, you learned that the current temperature there was -23 Celsius (-23°C). After the phone conversation, you wanted to convert this temperature to Fahrenheit degrees, °F, but you could not find a reference with the correct formulas. You then remembered that the relationship between °F and °C is linear. [UW]

   a. Using this and the knowledge that 32°F = 0 °C and 212 °F = 100 °C, find an equation that computes Celsius temperature in terms of Fahrenheit; i.e. an equation of the form C = "an expression involving only the variable F."

   b. Likewise, find an equation that computes Fahrenheit temperature in terms of Celsius temperature; i.e. an equation of the form F = "an expression involving only the variable C."

   c. How cold was it in Oslo in °F?

Section 2.2 Graphs of Linear Functions

When we are working with a new function, it is useful to know as much as we can about the function: its graph, where the function is zero, and any other special behaviors of the function. We will begin this exploration of linear functions with a look at graphs.

When graphing a linear function, there are three basic ways to graph it:
1) By plotting points (at least 2) and drawing a line through the points
2) Using the initial value (output when $x = 0$) and rate of change (slope)
3) Using transformations of the identity function $f(x) = x$

Example 1

Graph $f(x) = 5 - \dfrac{2}{3}x$ by plotting points

In general, we evaluate the function at two or more inputs to find at least two points on the graph. Usually it is best to pick input values that will "work nicely" in the equation. In this equation, multiples of 3 will work nicely due to the $\dfrac{2}{3}$ in the equation, and of course using $x = 0$ to get the vertical intercept. Evaluating $f(x)$ at $x = 0$, 3 and 6:

$$f(0) = 5 - \frac{2}{3}(0) = 5$$

$$f(3) = 5 - \frac{2}{3}(3) = 3$$

$$f(6) = 5 - \frac{2}{3}(6) = 1$$

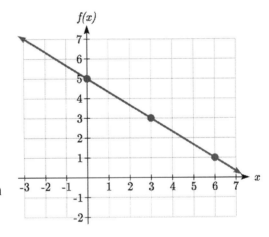

These evaluations tell us that the points $(0,5)$, $(3,3)$, and $(6,1)$ lie on the graph of the line. Plotting these points and drawing a line through them gives us the graph.

When using the initial value and rate of change to graph, we need to consider the graphical interpretation of these values. Remember the initial value of the function is the output when the input is zero, so in the equation $f(x) = b + mx$, the graph includes the point $(0, b)$. On the graph, this is the vertical intercept – the point where the graph crosses the vertical axis.

For the rate of change, it is helpful to recall that we calculated this value as

$$m = \frac{\text{change of output}}{\text{change of input}}$$

From a graph of a line, this tells us that if we divide the vertical difference, or rise, of the function outputs by the horizontal difference, or run, of the inputs, we will obtain the rate of change, also called slope of the line.

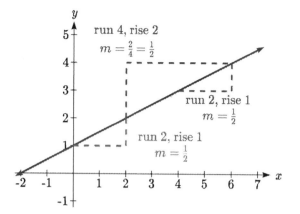

$$m = \frac{\text{change of output}}{\text{change of input}} = \frac{rise}{run}$$

Notice that this ratio is the same regardless of which two points we use.

Graphical Interpretation of a Linear Equation

Graphically, in the equation $f(x) = b + mx$,

b is the **vertical intercept** of the graph and tells us we can start our graph at $(0, b)$

m is the **slope of the line** and tells us how far to rise & run to get to the next point

Once we have at least 2 points, we can extend the graph of the line to the left and right.

Example 2

Graph $f(x) = 5 - \dfrac{2}{3}x$ using the vertical intercept and slope.

The vertical intercept of the function is $(0, 5)$, giving us a point on the graph of the line.

The slope is $-\dfrac{2}{3}$. This tells us that for every 3 units the graph "runs" in the horizontal, the vertical "rise" decreases by 2 units.

In graphing, we can use this by first plotting our vertical intercept on the graph, then using the slope to find a second point. From the initial value $(0, 5)$ the slope tells us that if we move to the right 3, we will move down 2, moving us to the point $(3, 3)$. We can continue this again to find a third point at $(6, 1)$. Finally, extend the line to the left and right, containing these points.

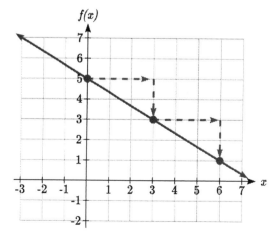

Try it Now

1. Consider that the slope $-\dfrac{2}{3}$ could also be written as $\dfrac{2}{-3}$. Using $\dfrac{2}{-3}$, find another point on the graph that has a negative x value.

Another option for graphing is to use transformations of the identity function $f(x) = x$.

In the equation $f(x) = mx$, the m is acting as the vertical stretch of the identity function. When m is negative, there is also a vertical reflection of the graph. Looking at some examples:

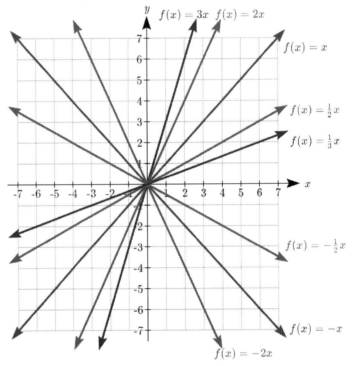

In $f(x) = mx + b$, the b acts as the vertical shift, moving the graph up and down without affecting the slope of the line. Some examples:

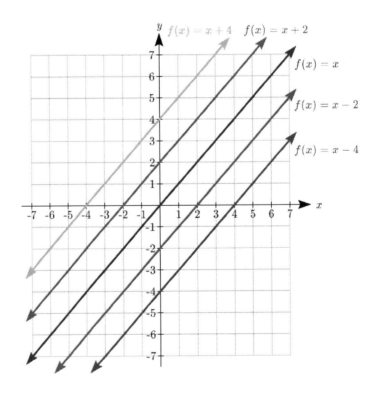

Using Vertical Stretches or Compressions along with Vertical Shifts is another way to look at identifying different types of linear functions. Although this may not be the easiest way for you to graph this type of function, make sure you practice each method.

Example 3

Graph $f(x) = -3 + \dfrac{1}{2}x$ using transformations.

The equation is the graph of the identity function vertically compressed by ½ and vertically shifted down 3.

Vertical compression   combined with Vertical shift

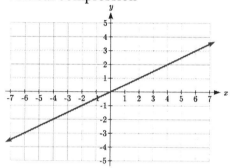   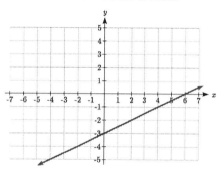

Notice how this nicely compares to the other method where the vertical intercept is found at (0, -3) and to get to another point we rise (go up vertically) by 1 unit and run (go horizontally) by 2 units to get to the next point (2, -2), and the next one (4, -1). In these three points (0, -3), (2, -2), and (4, -1), the output values change by +1, and the x values change by +2, corresponding with the slope $m = ½$.

Example 4

Match each equation with one of the lines in the graph below

$f(x) = 2x + 3$

$g(x) = 2x - 3$

$h(x) = -2x + 3$

$j(x) = \dfrac{1}{2}x + 3$

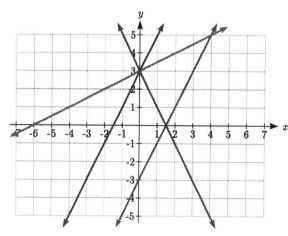

Only one graph has a vertical intercept of -3, so we can immediately match that graph with *g(x)*.

For the three graphs with a vertical
intercept at 3, only one has a negative
slope, so we can match that line with *h(x)*.
Of the other two, the steeper line would
have a larger slope, so we can match that
graph with equation *f(x)*, and the flatter
line with the equation *j(x)*.

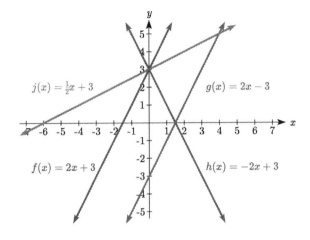

In addition to understanding the basic behavior of a linear function (increasing or
decreasing, recognizing the slope and vertical intercept), it is often helpful to know the
horizontal intercept of the function – where it crosses the horizontal axis.

Finding Horizontal Intercepts

The **horizontal intercept** of the function is where the graph crosses the horizontal
axis. If a function has a horizontal intercept, you can always find it by solving
$f(x) = 0$.

Example 5

Find the horizontal intercept of $f(x) = -3 + \dfrac{1}{2}x$

Setting the function equal to zero to find what input will put us on the horizontal axis,

$$0 = -3 + \frac{1}{2}x$$

$$3 = \frac{1}{2}x$$

$$x = 6$$

The graph crosses the horizontal axis at (6,0)

There are two special cases of lines: a horizontal line and a vertical line. In a horizontal line like the one graphed to the right, notice that between any two points, the change in the outputs is 0. In the slope equation, the numerator will be 0, resulting in a slope of 0. Using a slope of 0 in the $f(x) = b + mx$, the equation simplifies to $f(x) = b$.

Notice a horizontal line has a vertical intercept, but no horizontal intercept (unless it's the line $f(x) = 0$).

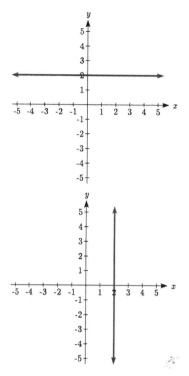

In the case of a vertical line, notice that between any two points, the change in the inputs is zero. In the slope equation, the denominator will be zero, and you may recall that we cannot divide by the zero; the slope of a vertical line is undefined. You might also notice that a vertical line is not a function. To write the equation of vertical line, we simply write input=value, like $x = b$.

Notice a vertical line has a horizontal intercept, but no vertical intercept (unless it's the line $x = 0$).

Horizontal and Vertical Lines
Horizontal lines have equations of the form $f(x) = b$
Vertical lines have equations of the form $x = a$

Example 6

Write an equation for the horizontal line graphed above.

This line would have equation $f(x) = 2$

Example 7

Write an equation for the vertical line graphed above.

This line would have equation $x = 2$

Try it Now

2. Describe the function $f(x) = 6 - 3x$ in terms of transformations of the identity function and find its horizontal intercept.

Parallel and Perpendicular Lines

When two lines are graphed together, the lines will be **parallel** if they are increasing at the same rate – if the rates of change are the same. In this case, the graphs will never cross (unless they're the same line).

Parallel Lines
Two lines are **parallel** if the slopes are equal (or, if both lines are vertical). In other words, given two linear equations $f(x) = b + m_1 x$ and $g(x) = b + m_2 x$, the lines will be parallel if $m_1 = m_2$.

Example 8
Find a line parallel to $f(x) = 6 + 3x$ that passes through the point (3, 0) We know the line we're looking for will have the same slope as the given line, $m = 3$. Using this and the given point, we can solve for the new line's vertical intercept: $g(x) = b + 3x$ $\qquad$ then at (3, 0), $0 = b + 3(3)$ $b = -9$ The line we're looking for is $g(x) = -9 + 3x$

If two lines are not parallel, one other interesting possibility is that the lines are perpendicular, which means the lines form a right angle (90 degree angle – a square corner) where they meet. In this case, the slopes when multiplied together will equal -1. Solving for one slope leads us to the definition:

Perpendicular Lines
Given two linear equations $f(x) = b + m_1 x$ and $g(x) = b + m_2 x$ The lines will be **perpendicular** if $m_1 m_2 = -1$, and so $m_2 = \dfrac{-1}{m_1}$ We often say the slope of a perpendicular line is the "negative reciprocal" of the other line's slope.

Example 9

Find the slope of a line perpendicular to a line with:

a) a slope of 2.   b) a slope of -4.   c) a slope of $\dfrac{2}{3}$.

If the original line had slope 2, the perpendicular line's slope would be $m_2 = \dfrac{-1}{2}$

If the original line had slope -4, the perpendicular line's slope would be $m_2 = \dfrac{-1}{-4} = \dfrac{1}{4}$

If the original line had slope $\dfrac{2}{3}$, the perpendicular line's slope would be $m_2 = \dfrac{-1}{{}^{2}\!/_{3}} = \dfrac{-3}{2}$

Example 10

Find the equation of a line perpendicular to $f(x) = 6 + 3x$ and passing through the point $(3, 0)$

The original line has slope $m = 3$. The perpendicular line will have slope $m = \dfrac{-1}{3}$.

Using this and the given point, we can find the equation for the line.

$g(x) = b - \dfrac{1}{3}x$ then at $(3, 0)$,

$0 = b - \dfrac{1}{3}(3)$

$b = 1$

The line we're looking for is $g(x) = 1 - \dfrac{1}{3}x$.

Try it Now

3. Given the line $h(t) = -4 + 2t$, find an equation for the line passing through $(0, 0)$ that
   is: a) parallel to $h(t)$.   b) perpendicular to $h(t)$.

Example 12

A line passes through the points (-2, 6) and (4, 5). Find the equation of a perpendicular line that passes through the point (4, 5).

From the two given points on the reference line, we can calculate the slope of that line:

$m_1 = \dfrac{5 - 6}{4 - (-2)} = \dfrac{-1}{6}$

The perpendicular line will have slope

$$m_2 = \frac{-1}{-1/6} = 6$$

We can then solve for the vertical intercept that makes the line pass through the desired point:

$g(x) = b + 6x$ then at (4, 5),

$5 = b + 6(4)$

$b = -19$

Giving the line $g(x) = -19 + 6x$

Intersections of Lines

The graphs of two lines will intersect if they are not parallel. They will intersect at the point that satisfies both equations. To find this point when the equations are given as functions, we can solve for an input value so that $f(x) = g(x)$. In other words, we can set the formulas for the lines equal, and solve for the input that satisfies the equation.

Example 13

Find the intersection of the lines $h(t) = 3t - 4$ and $j(t) = 5 - t$

Setting $h(t) = j(t)$,

$3t - 4 = 5 - t$

$4t = 9$

$t = \dfrac{9}{4}$

This tells us the lines intersect when the input is $\dfrac{9}{4}$.

We can then find the output value of the intersection point by evaluating either function at this input

$$j\left(\frac{9}{4}\right) = 5 - \frac{9}{4} = \frac{11}{4}$$

These lines intersect at the point $\left(\dfrac{9}{4}, \dfrac{11}{4}\right)$. Looking at the graph, this result seems reasonable.

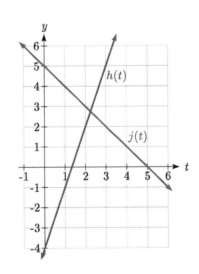

Two parallel lines can also intersect if they happen to be the same line. In that case, they intersect at every point on the lines.

Try it Now

4. Look at the graph in example 13 above and answer the following for the function $h(t)$:
   a. Vertical intercept coordinates
   b. Horizontal intercepts coordinates
   c. Slope
   d. Is $j(t)$ parallel or perpendicular to $h(t)$ (or neither)
   e. Is $h(t)$ an Increasing or Decreasing function (or neither)
   f. Write a transformation description from the identity toolkit function $f(x) = x$

Finding the intersection allows us to answer other questions as well, such as discovering when one function is larger than another.

Example 14

Using the functions from the previous example, for what values of t is $h(t) > j(t)$

To answer this question, it is helpful first to know where the functions are equal, since that is the point where $h(t)$ could switch from being greater to smaller than $j(t)$ or vice-versa. From the previous example, we know the functions are equal at $t = \dfrac{9}{4}$.

By examining the graph, we can see that $h(t)$, the function with positive slope, is going to be larger than the other function to the right of the intersection. So $h(t) > j(t)$ when $t > \dfrac{9}{4}$

Important Topics of this Section
Methods for graphing linear functions
Another name for slope = rise/run
Horizontal intercepts (a,0)
Horizontal lines
Vertical lines
Parallel lines
Perpendicular lines
Intersecting lines

Try it Now Answers

1. (-3,7) found by starting at the vertical intercept, going up 2 units and 3 in the negative horizontal direction. You could have also answered, (-6, 9) or (-9, 11) etc…

2. Vertically stretched by a factor of 3, Vertically flipped (flipped over the x axis), Vertically shifted up by 6 units.
   Horizontal intercept: $6-3x=0$ when $x=2$

3. Parallel $f(t) = 2t$; Perpendicular $g(t) = -\dfrac{1}{2}t$

4. Given $h(t) = 3t - 4$
   a. (0,-4)
   b. $\left(\dfrac{4}{3}, 0\right)$
   c. Slope 3
   d. Neither parallel nor perpendicular
   e. Increasing function
   f. Given the identity function, vertically stretch by 3 and shift down 4 units.

Section 2.2 Exercises

Match each linear equation with its graph

1. $f(x) = -x - 1$

2. $f(x) = -2x - 1$

3. $f(x) = -\dfrac{1}{2}x - 1$

4. $f(x) = 2$

5. $f(x) = 2 + x$

6. $f(x) = 3x + 2$

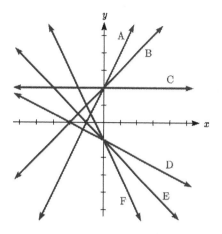

Sketch a line with the given features

7. An x-intercept of (-4, 0) and y-intercept of (0, -2)

8. An x-intercept of (-2, 0) and y-intercept of (0, 4)

9. A vertical intercept of (0, 7) and slope $-\dfrac{3}{2}$

10. A vertical intercept of (0, 3) and slope $\dfrac{2}{5}$

11. Passing through the points (-6,-2) and (6,-6)

12. Passing through the points (-3,-4) and (3,0)

Sketch the graph of each equation

13. $f(x) = -2x - 1$

14. $g(x) = -3x + 2$

15. $h(x) = \dfrac{1}{3}x + 2$

16. $k(x) = \dfrac{2}{3}x - 3$

17. $k(t) = 3 + 2t$

18. $p(t) = -2 + 3t$

19. $x = 3$

20. $x = -2$

21. $r(x) = 4$

22. $q(x) = 3$

23. If $g(x)$ is the transformation of $f(x) = x$ after a vertical compression by $3/4$, a shift left by 2, and a shift down by 4
    a. Write an equation for $g(x)$
    b. What is the slope of this line?
    c. Find the vertical intercept of this line.

24. If $g(x)$ is the transformation of $f(x) = x$ after a vertical compression by $1/3$, a shift right by 1, and a shift up by 3
    a. Write an equation for $g(x)$
    b. What is the slope of this line?
    c. Find the vertical intercept of this line.

Write the equation of the line shown

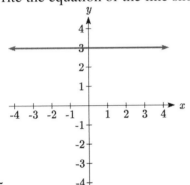

25.

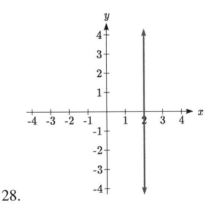

26.

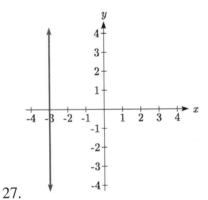

27.

28.

Find the horizontal and vertical intercepts of each equation

29. $f(x) = -x + 2$ 30. $g(x) = 2x + 4$

31. $h(x) = 3x - 5$ 32. $k(x) = -5x + 1$

33. $-2x + 5y = 20$ 34. $7x + 2y = 56$

Given below are descriptions of two lines. Find the slopes of Line 1 and Line 2. Is each pair of lines parallel, perpendicular or neither?

35. Line 1: Passes through $(0,6)$ and $(3,-24)$
    Line 2: Passes through $(-1,19)$ and $(8,-71)$

36. Line 1: Passes through $(-8,-55)$ and $(10,89)$
    Line 2: Passes through $(9,-44)$ and $(4,-14)$

37. Line 1: Passes through $(2,3)$ and $(4,-1)$
    Line 2: Passes through $(6,3)$ and $(8,5)$

38. Line 1: Passes through $(1,7)$ and $(5,5)$
    Line 2: Passes through $(-1,-3)$ and $(1,1)$

39. Line 1: Passes through $(0,5)$ and $(3,3)$
    Line 2: Passes through $(1,-5)$ and $(3,-2)$

40. Line 1: Passes through $(2,5)$ and $(5,-1)$
    Line 2: Passes through $(-3,7)$ and $(3,-5)$

41. Write an equation for a line parallel to $f(x) = -5x - 3$ and passing through the point $(2,-12)$

42. Write an equation for a line parallel to $g(x) = 3x - 1$ and passing through the point $(4,9)$

43. Write an equation for a line perpendicular to $h(t) = -2t + 4$ and passing through the point $(-4,-1)$

44. Write an equation for a line perpendicular to $p(t) = 3t + 4$ and passing through the point $(3,1)$

45. Find the point at which the line $f(x) = -2x - 1$ intersects the line $g(x) = -x$

46. Find the point at which the line $f(x) = 2x + 5$ intersects the line $g(x) = -3x - 5$

47. Use algebra to find the point at which the line $f(x) = -\dfrac{4}{5}x + \dfrac{274}{25}$ intersects the line

$h(x) = \dfrac{9}{4}x + \dfrac{73}{10}$

48. Use algebra to find the point at which the line $f(x) = \dfrac{7}{4}x + \dfrac{457}{60}$ intersects the line

$g(x) = \dfrac{4}{3}x + \dfrac{31}{5}$

49. A car rental company offers two plans for renting a car.
    Plan A: 30 dollars per day and 18 cents per mile
    Plan B: 50 dollars per day with free unlimited mileage
    How many miles would you need to drive for plan B to save you money?

50. You're comparing two cell phone companies.
    Company A: $20/month for unlimited talk and text, and $10/GB for data.
    Company B: $65/month for unlimited talk, text, and data.
    Under what circumstances will company A save you money?

Find a formula for each piecewise defined function.

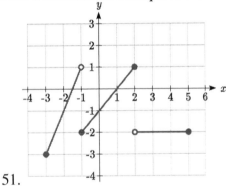

51.

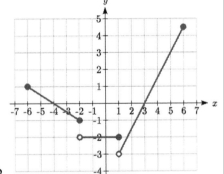

52.

53. Sketch an accurate picture of the line having equation $f(x) = 2 - \dfrac{1}{2}x$. Let c be an

    unknown constant. [UW]
    a. Find the point of intersection between the line you have graphed and the
       line $g(x) = 1 + cx$; your answer will be a point in the xy plane whose
       coordinates involve the unknown c.
    b. Find c so that the intersection point in (a) has x-coordinate 10.
    c. Find c so that the intersection point in (a) lies on the x-axis.

Section 2.3 Modeling with Linear Functions

When modeling scenarios with a linear function and solving problems involving quantities changing linearly, we typically follow the same problem solving strategies that we would use for any type of function:

Problem solving strategy

1) Identify changing quantities, and then carefully and clearly define descriptive variables to represent those quantities. When appropriate, sketch a picture or define a coordinate system.

2) Carefully read the problem to identify important information. Look for information giving values for the variables, or values for parts of the functional model, like slope and initial value.

3) Carefully read the problem to identify what we are trying to find, identify, solve, or interpret.

4) Identify a solution pathway from the provided information to what we are trying to find. Often this will involve checking and tracking units, building a table or even finding a formula for the function being used to model the problem.

5) When needed, find a formula for the function.

6) Solve or evaluate using the formula you found for the desired quantities.

7) Reflect on whether your answer is reasonable for the given situation and whether it makes sense mathematically.

8) Clearly convey your result using appropriate units, and answer in full sentences when appropriate.

Example 1

Emily saved up $3500 for her summer visit to Seattle. She anticipates spending $400 each week on rent, food, and fun. Find and interpret the horizontal intercept and determine a reasonable domain and range for this function.

In the problem, there are two changing quantities: time and money. The amount of money she has remaining while on vacation depends on how long she stays. We can define our variables, including units.
Output: M, money remaining, in dollars
Input: t, time, in weeks

Reading the problem, we identify two important values. The first, $3500, is the initial value for M. The other value appears to be a rate of change – the units of dollars per week match the units of our output variable divided by our input variable. She is spending money each week, so you should recognize that the amount of money remaining is decreasing each week and the slope is negative.

To answer the first question, looking for the horizontal intercept, it would be helpful to have an equation modeling this scenario. Using the intercept and slope provided in the problem, we can write the equation: $M(t) = 3500 - 400t$.

To find the horizontal intercept, we set the output to zero, and solve for the input:
$0 = 3500 - 400t$

$t = \dfrac{3500}{400} = 8.75$

The horizontal intercept is 8.75 weeks. Since this represents the input value where the output will be zero, interpreting this, we could say: Emily will have no money left after 8.75 weeks.

When modeling any real life scenario with functions, there is typically a limited domain over which that model will be valid – almost no trend continues indefinitely. In this case, it certainly doesn't make sense to talk about input values less than zero. It is also likely that this model is not valid after the horizontal intercept (unless Emily's going to start using a credit card and go into debt).

The domain represents the set of input values and so the reasonable domain for this function is $0 \le t \le 8.75$.

However, in a real world scenario, the rental might be weekly or nightly. She may not be able to stay a partial week and so all options should be considered. Emily could stay in Seattle for 0 to 8 full weeks (and a couple of days), but would have to go into debt to stay 9 full weeks, so restricted to whole weeks, a reasonable domain without going in to debt would be $0 \le t \le 8$, or $0 \le t \le 9$ if she went into debt to finish out the last week.

The range represents the set of output values and she starts with $3500 and ends with $0 after 8.75 weeks so the corresponding range is $0 \le M(t) \le 3500$. If we limit the rental to whole weeks, however, the range would change. If she left after 8 weeks because she didn't have enough to stay for a full 9 weeks, she would have $M(8) = 3500-400(8) =$ $300 dollars left after 8 weeks, giving a range of $300 \le M(t) \le 3500$. If she wanted to stay the full 9 weeks she would be $100 in debt giving a range of $-100 \le M(t) \le 3500$.

Most importantly remember that domain and range are tied together, and what ever you decide is most appropriate for the domain (the independent variable) will dictate the requirements for the range (the dependent variable).

Try it Now

1. A database manager is loading a large table from backups. Getting impatient, she notices 1.2 million rows had been loaded. Ten minutes later, 2.5 million rows had been loaded. How much longer will she have to wait for all 80 million rows to load?

Example 2

Jamal is choosing between two moving companies. The first, U-Haul, charges an up-front fee of $20, then 59 cents a mile. The second, Budget, charges an up-front fee of $16, then 63 cents a mile[3]. When will U-Haul be the better choice for Jamal?

The two important quantities in this problem are the cost, and the number of miles that are driven. Since we have two companies to consider, we will define two functions:

Input: m, miles driven
Outputs:
$Y(m)$: cost, in dollars, for renting from U-Haul
$B(m)$: cost, in dollars, for renting from Budget

Reading the problem carefully, it appears that we were given an initial cost and a rate of change for each company. Since our outputs are measured in dollars but the costs per mile given in the problem are in cents, we will need to convert these quantities to match our desired units: $0.59 a mile for U-Haul, and $0.63 a mile for Budget.

Looking to what we're trying to find, we want to know when U-Haul will be the better choice. Since all we have to make that decision from is the costs, we are looking for when U-Haul will cost less, or when $Y(m) < B(m)$. The solution pathway will lead us to find the equations for the two functions, find the intersection, then look to see where the $Y(m)$ function is smaller. Using the rates of change and initial charges, we can write the equations:

$Y(m) = 20 + 0.59m$

$B(m) = 16 + 0.63m$

These graphs are sketched to the right, with $Y(m)$ drawn dashed.

To find the intersection, we set the equations equal and solve:
$Y(m) = B(m)$
$20 + 0.59m = 16 + 0.63m$

$4 = 0.04m$

$m = 100$

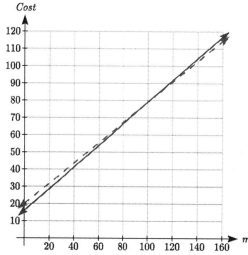

This tells us that the cost from the two companies will be the same if 100 miles are driven. Either by looking at the graph, or noting that $Y(m)$ is growing at a slower rate, we can conclude that U-Haul will be the cheaper price when more than 100 miles are driven.

---

[3] Rates retrieved Aug 2, 2010 from http://www.budgettruck.com and http://www.uhaul.com/

Example 3

A town's population has been growing linearly. In 2004 the population was 6,200. By 2009 the population had grown to 8,100. If this trend continues,
a. Predict the population in 2013
b. When will the population reach 15000?

The two changing quantities are the population and time. While we could use the actual year value as the input quantity, doing so tends to lead to very ugly equations, since the vertical intercept would correspond to the year 0, more than 2000 years ago!
To make things a little nicer, and to make our lives easier too, we will define our input as years since 2004:
Input: t, years since 2004
Output: $P(t)$, the town's population

The problem gives us two input-output pairs. Converting them to match our defined variables, the year 2004 would correspond to $t = 0$, giving the point (0, 6200). Notice that through our clever choice of variable definition, we have "given" ourselves the vertical intercept of the function. The year 2009 would correspond to $t = 5$, giving the point (5, 8100).

To predict the population in 2013 ($t = 9$), we would need an equation for the population. Likewise, to find when the population would reach 15000, we would need to solve for the input that would provide an output of 15000. Either way, we need an equation. To find it, we start by calculating the rate of change:

$$m = \frac{8100 - 6200}{5 - 0} = \frac{1900}{5} = 380 \text{ people per year}$$

Since we already know the vertical intercept of the line, we can immediately write the equation:
$$P(t) = 6200 + 380t$$

To predict the population in 2013, we evaluate our function at $t = 9$
$$P(9) = 6200 + 380(9) = 9620$$
If the trend continues, our model predicts a population of 9,620 in 2013.

To find when the population will reach 15,000, we can set $P(t) = 15000$ and solve for t.
$$15000 = 6200 + 380t$$

$$8800 = 380t$$

$$t \approx 23.158$$

Our model predicts the population will reach 15,000 in a little more than 23 years after 2004, or somewhere around the year 2027.

Example 4

Anna and Emanuel start at the same intersection. Anna walks east at 4 miles per hour while Emanuel walks south at 3 miles per hour. They are communicating with a two-way radio with a range of 2 miles. How long after they start walking will they fall out of radio contact?

In essence, we can partially answer this question by saying they will fall out of radio contact when they are 2 miles apart, which leads us to ask a new question: how long will it take them to be 2 miles apart?

In this problem, our changing quantities are time and the two peoples' positions, but ultimately we need to know how long will it take for them to be 2 miles apart. We can see that time will be our input variable, so we'll define
Input: t, time in hours.

Since it is not obvious how to define our output variables, we'll start by drawing a picture.

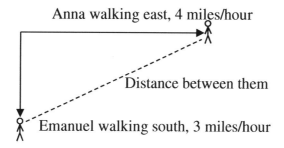

Anna walking east, 4 miles/hour

Distance between them

Emanuel walking south, 3 miles/hour

Because of the complexity of this question, it may be helpful to introduce some intermediary variables. These are quantities that we aren't directly interested in, but seem important to the problem. For this problem, Anna's and Emanuel's distances from the starting point seem important. To notate these, we are going to define a coordinate system, putting the "starting point" at the intersection where they both started, then we're going to introduce a variable, A, to represent Anna's position, and define it to be a measurement from the starting point, in the eastward direction. Likewise, we'll introduce a variable, E, to represent Emanuel's position, measured from the starting point in the southward direction. Note that in defining the coordinate system we specified both the origin, or starting point, of the measurement, as well as the direction of measure.

While we're at it, we'll define a third variable, D, to be the measurement of the distance between Anna and Emanuel. Showing the variables on the picture is often helpful:
Looking at the variables on the picture, we remember we need to know how long it takes for D, the distance between them, to equal 2 miles.

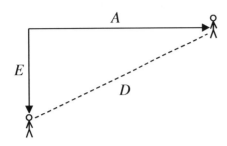

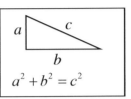

Seeing this picture we remember that in order to find the distance between the two, we can use the Pythagorean Theorem, a property of right triangles.

From here, we can now look back at the problem for relevant information. Anna is walking 4 miles per hour, and Emanuel is walking 3 miles per hour, which are rates of change. Using those, we can write formulas for the distance each has walked.

They both start at the same intersection and so when $t = 0$, the distance travelled by each person should also be 0, so given the rate for each, and the initial value for each, we get:

$A(t) = 4t$

$E(t) = 3t$

Using the Pythagorean theorem we get:

$D(t)^2 = A(t)^2 + E(t)^2$ substitute in the function formulas

$D(t)^2 = (4t)^2 + (3t)^2 = 16t^2 + 9t^2 = 25t^2$ solve for $D(t)$ using the square root

$D(t) = \pm\sqrt{25t^2} = \pm 5|t|$

Since in this scenario we are only considering positive values of t and our distance $D(t)$ will always be positive, we can simplify this answer to $D(t) = 5t$

Interestingly, the distance between them is also a linear function. Using it, we can now answer the question of when the distance between them will reach 2 miles:

$D(t) = 2$

$5t = 2$

$t = \dfrac{2}{5} = 0.4$

They will fall out of radio contact in 0.4 hours, or 24 minutes.

Example 5

There is currently a straight road leading from the town of Westborough to a town 30 miles east and 10 miles north. Partway down this road, it junctions with a second road, perpendicular to the first, leading to the town of Eastborough. If the town of Eastborough is located 20 miles directly east of the town of Westborough, how far is the road junction from Westborough?

It might help here to draw a picture of the situation. It would then be helpful to introduce a coordinate system. While we could place the origin anywhere, placing it at Westborough seems convenient. This puts the other town at coordinates (30, 10), and Eastborough at (20, 0).

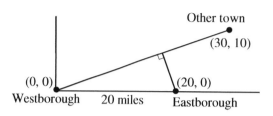

Using this point along with the origin, we can find the slope of the line from Westborough to the other town: $m = \dfrac{10-0}{30-0} = \dfrac{1}{3}$. This gives the equation of the road from Westborough to the other town to be $W(x) = \dfrac{1}{3}x$.

From this, we can determine the perpendicular road to Eastborough will have slope $m = -3$. Since the town of Eastborough is at the point (20, 0), we can find the equation:

$E(x) = -3x + b$ plug in the point (20, 0)

$0 = -3(20) + b$

$b = 60$

$E(x) = -3x + 60$

We can now find the coordinates of the junction of the roads by finding the intersection of these lines. Setting them equal,

$\dfrac{1}{3}x = -3x + 60$

$\dfrac{10}{3}x = 60$

$10x = 180$

$x = 18$ Substituting this back into $W(x)$

$y = W(18) = \dfrac{1}{3}(18) = 6$

The roads intersect at the point (18, 6). Using the distance formula, we can now find the distance from Westborough to the junction:

$dist = \sqrt{(18-0)^2 + (6-0)^2} \approx 18.934$ miles.

Important Topics of this Section

The problem solving process

1) Identify changing quantities, and then carefully and clearly define descriptive variables to represent those quantities. When appropriate, sketch a picture or define a coordinate system.

2) Carefully read the problem to identify important information. Look for information giving values for the variables, or values for parts of the functional model, like slope and initial value.

3) Carefully read the problem to identify what we are trying to find, identify, solve, or interpret.

4) Identify a solution pathway from the provided information to what we are trying to find. Often this will involve checking and tracking units, building a table or even finding a formula for the function being used to model the problem.

5) When needed, find a formula for the function.

6) Solve or evaluate using the formula you found for the desired quantities.

7) Reflect on whether your answer is reasonable for the given situation and whether it makes sense mathematically.

8) Clearly convey your result using appropriate units, and answer in full sentences when appropriate.

Try it Now

1. Letting t be the number of minutes since she got impatient, and N be the number rows loaded, in millions, we have two points: (0, 1.2) and (10, 2.5).

The slope is $m = \dfrac{2.5-1.2}{10-0} = \dfrac{1.3}{10} = 0.13$ million rows per minute.

We know the N intercept, so we can write the equation:
$N = 0.13t + 1.2$

To determine how long she will have to wait, we need to solve for when $N = 80$.
$N = 0.13t + 1.2 = 80$
$0.13t = 78.8$
$t = \dfrac{78.8}{0.13} \approx 606$. She'll have to wait another 606 minutes, about 10 hours.

Section 2.3 Exercises

1. In 2004, a school population was 1001. By 2008 the population had grown to 1697. Assume the population is changing linearly.
   a. How much did the population grow between the year 2004 and 2008?
   b. How long did it take the population to grow from 1001 students to 1697 students?
   c. What is the average population growth per year?
   d. What was the population in the year 2000?
   e. Find an equation for the population, P, of the school t years after 2000.
   f. Using your equation, predict the population of the school in 2011.

2. In 2003, a town's population was 1431. By 2007 the population had grown to 2134. Assume the population is changing linearly.
   a. How much did the population grow between the year 2003 and 2007?
   b. How long did it take the population to grow from 1431 people to 2134?
   c. What is the average population growth per year?
   d. What was the population in the year 2000?
   e. Find an equation for the population, P, of the town t years after 2000.
   f. Using your equation, predict the population of the town in 2014.

3. A phone company has a monthly cellular plan where a customer pays a flat monthly fee and then a certain amount of money per minute used on the phone. If a customer uses 410 minutes, the monthly cost will be $71.50. If the customer uses 720 minutes, the monthly cost will be $118.
   a. Find a linear equation for the monthly cost of the cell plan as a function of x, the number of monthly minutes used.
   b. Interpret the slope and vertical intercept of the equation.
   c. Use your equation to find the total monthly cost if 687 minutes are used.

4. A phone company has a monthly cellular data plan where a customer pays a flat monthly fee and then a certain amount of money per megabyte (MB) of data used on the phone. If a customer uses 20 MB, the monthly cost will be $11.20. If the customer uses 130 MB, the monthly cost will be $17.80.
   a. Find a linear equation for the monthly cost of the data plan as a function of x, the number of MB used.
   b. Interpret the slope and vertical intercept of the equation.
   c. Use your equation to find the total monthly cost if 250 MB are used.

5. In 1991, the moose population in a park was measured to be 4360. By 1999, the population was measured again to be 5880. If the population continues to change linearly,
    a. Find a formula for the moose population, P.
    b. What does your model predict the moose population to be in 2003?

6. In 2003, the owl population in a park was measured to be 340. By 2007, the population was measured again to be 285. If the population continues to change linearly,
    a. Find a formula for the owl population, P.
    b. What does your model predict the owl population to be in 2012?

7. The Federal Helium Reserve held about 16 billion cubic feet of helium in 2010, and is being depleted by about 2.1 billion cubic feet each year.
    a. Give a linear equation for the remaining federal helium reserves, R, in terms of t, the number of years since 2010.
    b. In 2015, what will the helium reserves be?
    c. If the rate of depletion doesn't change, when will the Federal Helium Reserve be depleted?

8. Suppose the world's current oil reserves are 1820 billion barrels. If, on average, the total reserves is decreasing by 25 billion barrels of oil each year:
    a. Give a linear equation for the remaining oil reserves, R, in terms of t, the number of years since now.
    b. Seven years from now, what will the oil reserves be?
    c. If the rate of depletion isn't change, when will the world's oil reserves be depleted?

9. You are choosing between two different prepaid cell phone plans. The first plan charges a rate of 26 cents per minute. The second plan charges a monthly fee of $19.95 *plus* 11 cents per minute. How many minutes would you have to use in a month in order for the second plan to be preferable?

10. You are choosing between two different window washing companies. The first charges $5 per window. The second charges a base fee of $40 plus $3 per window. How many windows would you need to have for the second company to be preferable?

11. When hired at a new job selling jewelry, you are given two pay options:
Option A: Base salary of $17,000 a year, with a commission of 12% of your sales
Option B: Base salary of $20,000 a year, with a commission of 5% of your sales
How much jewelry would you need to sell for option A to produce a larger income?

12. When hired at a new job selling electronics, you are given two pay options:
    Option A: Base salary of $14,000 a year, with a commission of 10% of your sales
    Option B: Base salary of $19,000 a year, with a commission of 4% of your sales
    How much electronics would you need to sell for option A to produce a larger income?

13. Find the area of a triangle bounded by the y axis, the line $f(x) = 9 - \dfrac{6}{7}x$, and the line perpendicular to $f(x)$ that passes through the origin.

14. Find the area of a triangle bounded by the x axis, the line $f(x) = 12 - \dfrac{1}{3}x$, and the line perpendicular to $f(x)$ that passes through the origin.

15. Find the area of a parallelogram bounded by the y axis, the line $x = 3$, the line $f(x) = 1 + 2x$, and the line parallel to $f(x)$ passing through (2, 7)

16. Find the area of a parallelogram bounded by the x axis, the line $g(x) = 2$, the line $f(x) = 3x$, and the line parallel to $f(x)$ passing through (6, 1)

17. If $b > 0$ and $m < 0$, then the line $f(x) = b + mx$ cuts off a triangle from the first quadrant. Express the area of that triangle in terms of m and b.   [UW]

18. Find the value of m so the lines $f(x) = mx + 5$ and $g(x) = x$ and the y-axis form a triangle with an area of 10.   [UW]

19. The median home values in Mississippi and Hawaii (adjusted for inflation) are shown below. If we assume that the house values are changing linearly,

Year	Mississippi	Hawaii
1950	25200	74400
2000	71400	272700

   a. In which state have home values increased at a higher rate?
   b. If these trends were to continue, what would be the median home value in Mississippi in 2010?
   c. If we assume the linear trend existed before 1950 and continues after 2000, the two states' median house values will be (or were) equal in what year? (The answer might be absurd)

20. The median home value ins Indiana and Alabama (adjusted for inflation) are shown below. If we assume that the house values are changing linearly,

Year	Indiana	Alabama
1950	37700	27100
2000	94300	85100

    a. In which state have home values increased at a higher rate?

    b. If these trends were to continue, what would be the median home value in Indiana in 2010?

    c. If we assume the linear trend existed before 1950 and continues after 2000, the two states' median house values will be (or were) equal in what year? (The answer might be absurd)

21. Pam is taking a train from the town of Rome to the town of Florence. Rome is located 30 miles due West of the town of Paris. Florence is 25 miles East, and 45 miles North of Rome. On her trip, how close does Pam get to Paris? [UW]

22. You're flying from Joint Base Lewis-McChord (JBLM) to an undisclosed location 226 km south and 230 km east. Mt. Rainier is located approximately 56 km east and 40 km south of JBLM. If you are flying at a constant speed of 800 km/hr, how long after you depart JBLM will you be the closest to Mt. Rainier?

Section 2.4 Fitting Linear Models to Data

In the real world, rarely do things follow trends perfectly. When we expect the trend to behave linearly, or when inspection suggests the trend is behaving linearly, it is often desirable to find an equation to approximate the data. Finding an equation to approximate the data helps us understand the behavior of the data and allows us to use the linear model to make predictions about the data, inside and outside of the data range.

Example 1

The table below shows the number of cricket chirps in 15 seconds, and the air temperature, in degrees Fahrenheit[4]. Plot this data, and determine whether the data appears to be linearly related.

chirps	44	35	20.4	33	31	35	18.5	37	26
Temp	80.5	70.5	57	66	68	72	52	73.5	53

Plotting this data, it appears there may be a trend, and that the trend appears roughly linear, though certainly not perfectly so.

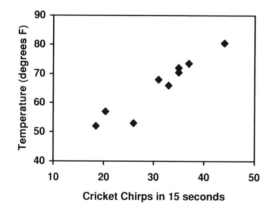

The simplest way to find an equation to approximate this data is to try to "eyeball" a line that seems to fit the data pretty well, then find an equation for that line based on the slope and intercept.

You can see from the trend in the data that the number of chirps increases as the temperature increases. As you consider a function for this data you should know that you are looking at an increasing function or a function with a positive slope.

---

[4] Selected data from http://classic.globe.gov/fsl/scientistsblog/2007/10/. Retrieved Aug 3, 2010

1. a. What descriptive variables would you choose to represent Temperature & Chirps?
   b. Which variable is the independent variable and which is the dependent variable?
   c. Based on this data and the graph, what is a reasonable domain & range?
   d. Based on the data alone, is this function one-to-one, explain?

Example 2

Using the table of values from the previous example, find a linear function that fits the data by "eyeballing" a line that seems to fit.

On a graph, we could try sketching in a line. Note the scale on the axes have been adjusted to start at zero to include the vertical axis and vertical intercept in the graph.

Using the starting and ending points of our "hand drawn" line, points (0, 30) and (50, 90), this graph has a slope of $m = \dfrac{60}{50} = 1.2$ and a vertical intercept at 30, giving an equation of

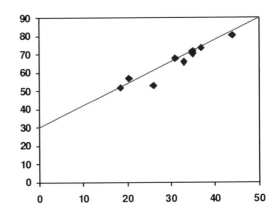

$T(c) = 30 + 1.2c$

where c is the number of chirps in 15 seconds, and $T(c)$ is the temperature in degrees Fahrenheit.

This linear equation can then be used to approximate the solution to various questions we might ask about the trend. While the data does not perfectly fall on the linear equation, the equation is our best guess as to how the relationship will behave outside of the values we have data for. There is a difference, though, between making predictions inside the domain and range of values we have data for, and outside that domain and range.

Interpolation and Extrapolation
Interpolation: When we predict a value inside the domain and range of the data
Extrapolation: When we predict a value outside the domain and range of the data

For the Temperature as a function of chirps in our hand drawn model above,
- Interpolation would occur if we used our model to predict temperature when the values for chirps are between 18.5 and 44.
- Extrapolation would occur if we used our model to predict temperature when the values for chirps are less than 18.5 or greater than 44.

Example 3

a) Would predicting the temperature when crickets are chirping 30 times in 15 seconds be interpolation or extrapolation? Make the prediction, and discuss if it is reasonable.

b) Would predicting the number of chirps crickets will make at 40 degrees be interpolation or extrapolation? Make the prediction, and discuss if it is reasonable.

With our cricket data, our number of chirps in the data provided varied from 18.5 to 44. A prediction at 30 chirps per 15 seconds is inside the domain of our data, so would be interpolation. Using our model:
$T(30) = 30 + 1.2(30) = 66$ degrees.

Based on the data we have, this value seems reasonable.

The temperature values varied from 52 to 80.5. Predicting the number of chirps at 40 degrees is extrapolation since 40 is outside the range of our data. Using our model:
$40 = 30 + 1.2c$

$10 = 1.2c$

$c \approx 8.33$

Our model predicts the crickets would chirp 8.33 times in 15 seconds. While this might be possible, we have no reason to believe our model is valid outside the domain and range. In fact, generally crickets stop chirping altogether below around 50 degrees.

When our model no longer applies after some point, it is sometimes called **model breakdown**.

Try it Now
1. What temperature would you predict if you counted 20 chirps in 15 seconds?

Fitting Lines with Technology

While eyeballing a line works reasonably well, there are statistical techniques for fitting a line to data that minimize the differences between the line and data values[5]. This technique is called **least-square regression**, and can be computed by many graphing calculators, spreadsheet software like Excel or Google Docs, statistical software, and many web-based calculators[6].

---

[5] Technically, the method minimizes the sum of the squared differences in the vertical direction between the line and the data values.
[6] For example, http://www.shodor.org/unchem/math/lls/leastsq.html

Example 4

Find the least-squares regression line using the cricket chirp data from above.

Using the cricket chirp data from earlier, with technology we obtain the equation:
$$T(c) = 30.281 + 1.143c$$

Notice that this line is quite similar to the equation we "eyeballed", but should fit the data better. Notice also that using this equation would change our prediction for the temperature when hearing 30 chirps in 15 seconds from 66 degrees to:
$$T(30) = 30.281 + 1.143(30) = 64.571 \approx 64.6$$
degrees.

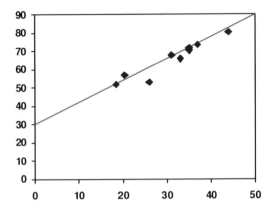

Most calculators and computer software will also provide you with the **correlation coefficient**, a measure of how closely the line fits the data.

Correlation Coefficient

The **correlation coefficient** is a value, r, between -1 and 1.

$r > 0$ suggests a positive (increasing) relationship

$r < 0$ suggests a negative (decreasing) relationship

The closer the value is to 0, the more scattered the data

The closer the value is to 1 or -1, the less scattered the data is

---

The correlation coefficient provides an easy way to get some idea of how close to a line the data falls.

We should only compute the correlation coefficient for data that follows a linear pattern; if the data exhibits a non-linear pattern, the correlation coefficient is meaningless. To get a sense for the relationship between the value of r and the graph of the data, here are some large data sets with their correlation coefficients:

Examples of Correlation Coefficient Values

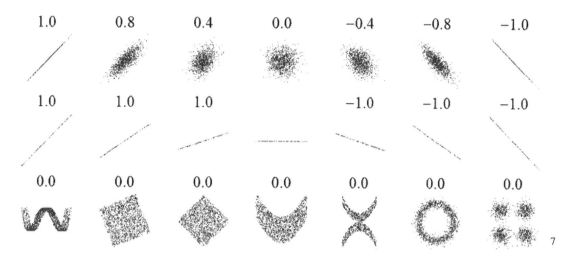

Example 5

Calculate the correlation coefficient for our cricket data.

Because the data appears to follow a linear pattern, we can use technology to calculate $r = 0.9509$. Since this value is very close to 1, it suggests a strong increasing linear relationship.

Example 6

Gasoline consumption in the US has been increasing steadily. Consumption data from 1994 to 2004 is shown below.[8]   Determine if the trend is linear, and if so, find a model for the data. Use the model to predict the consumption in 2008.

Year	'94	'95	'96	'97	'98	'99	'00	'01	'02	'03	'04
Consumption (billions of gallons)	113	116	118	119	123	125	126	128	131	133	136

To make things simpler, a new input variable is introduced, t, representing years since 1994.

Using technology, the correlation coefficient was calculated to be 0.9965, suggesting a very strong increasing linear trend.

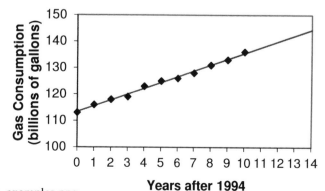

Years after 1994

[7] http://en.wikipedia.org/wiki/File:Correlation_examples.png
[8] http://www.bts.gov/publications/national_transportation_statistics/2005/html/table_04_10.html

The least-squares regression equation is:
$C(t) = 113.318 + 2.209t$.

Using this to predict consumption in 2008 ($t = 14$),
$C(14) = 113.318 + 2.209(14) = 144.244$ billions of gallons

The model predicts 144.244 billion gallons of gasoline will be consumed in 2008.

Try it Now

2. Use the model created by technology in example 6 to predict the gas consumption in 2011. Is this an interpolation or an extrapolation?

Important Topics of this Section

Fitting linear models to data by hand

Fitting linear models to data using technology

Interpolation

Extrapolation

Correlation coefficient

Flashback Answers

1. a. T = Temperature, C = Chirps (answers may vary)
   b. Independent (Chirps), Dependent (Temperature)
   c. Reasonable Domain (18.5, 44), Reasonable Range (52, 80.5) (answers may vary)
   d. NO, it is not one-to-one, there are two different output values for 35 chirps.

Try it Now Answers

1. 54 degrees Fahrenheit
2. 150.871 billion gallons; extrapolation

Section 2.4 Exercises

1. The following is data for the first and second quiz scores for 8 students in a class. Plot the points, then sketch a line that fits the data.

First Quiz	11	20	24	25	33	42	46	49
Second Quiz	10	16	23	28	30	39	40	49

2. Eight students were asked to estimate their score on a 10 point quiz. Their estimated and actual scores are given. Plot the points, then sketch a line that fits the data.

Predicted	5	7	6	8	10	9	10	7
Actual	6	6	7	8	9	9	10	6

Based on each set of data given, calculate the regression line using your calculator or other technology tool, and determine the correlation coefficient.

3.

x	y
5	4
7	12
10	17
12	22
15	24

4.

x	y
8	23
15	41
26	53
31	72
56	103

5.

x	y
3	21.9
4	22.22
5	22.74
6	22.26
7	20.78
8	17.6
9	16.52
10	18.54
11	15.76
12	13.68
13	14.1
14	14.02
15	11.94
16	12.76
17	11.28
18	9.1

6.

x	y
4	44.8
5	43.1
6	38.8
7	39
8	38
9	32.7
10	30.1
11	29.3
12	27
13	25.8
14	24.7
15	22
16	20.1
17	19.8
18	16.8

7. A regression was run to determine if there is a relationship between hours of TV watched per day (x) and number of situps a person can do (y). The results of the regression are given below. Use this to predict the number of situps a person who watches 11 hours of TV can do.

```
y=ax+b
a=-1.341
b=32.234
r²=0.803
r=-0.896
```

8. A regression was run to determine if there is a relationship between the diameter of a tree (x, in inches) and the tree's age (y, in years). The results of the regression are given below. Use this to predict the age of a tree with diameter 10 inches.

```
y=ax+b
a=6.301
b=-1.044
r²=0.940
r=-0.970
```

Match each scatterplot shown below with one of the four specified correlations.

9. $r = 0.95$ 10. $r = -0.89$ 11. $r = 0.26$ 12. $r = -0.39$

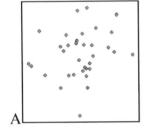
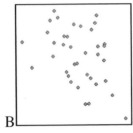
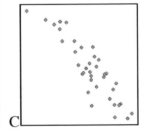
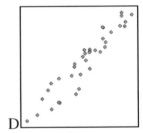

A   B   C   D

13. The US census tracks the percentage of persons 25 years or older who are college graduates. That data for several years is given below. Determine if the trend appears linear. If so and the trend continues, in what year will the percentage exceed 35%?

Year	1990	1992	1994	1996	1998	2000	2002	2004	2006	2008
Percent Graduates	21.3	21.4	22.2	23.6	24.4	25.6	26.7	27.7	28	29.4

14. The US import of wine (in hectoliters) for several years is given below. Determine if the trend appears linear. If so and the trend continues, in what year will imports exceed 12,000 hectoliters?

Year	1992	1994	1996	1998	2000	2002	2004	2006	2008	2009
Imports	2665	2688	3565	4129	4584	5655	6549	7950	8487	9462

Section 2.5 Absolute Value Functions

So far in this chapter we have been studying the behavior of linear functions. The Absolute Value Function is a piecewise-defined function made up of two linear functions. The name, Absolute Value Function, should be familiar to you from Section 1.2. In its basic form $f(x) = |x|$ it is one of our toolkit functions.

Absolute Value Function

The absolute value function can be defined as

$$f(x) = |x| = \begin{cases} x & if & x \geq 0 \\ -x & if & x < 0 \end{cases}$$

The absolute value function is commonly used to determine the distance between two numbers on the number line. Given two values a and b, then $|a - b|$ will give the distance, a positive quantity, between these values, regardless of which value is larger.

Example 1

Describe all values, x, within a distance of 4 from the number 5.

We want the distance between x and 5 to be less than or equal to 4. The distance can be represented using the absolute value, giving the expression

$|x - 5| \leq 4$

Example 2

A 2010 poll reported 78% of Americans believe that people who are gay should be able to serve in the US military, with a reported margin of error of 3%[9]. The margin of error tells us how far off the actual value could be from the survey value[10]. Express the set of possible values using absolute values.

Since we want the size of the difference between the actual percentage, p, and the reported percentage to be less than 3%,

$|p - 78| \leq 3$

---

[9] http://www.pollingreport.com/civil.htm, retrieved August 4, 2010
[10] Technically, margin of error usually means that the surveyors are 95% confident that actual value falls within this range.

Try it Now
1. Students who score within 20 points of 80 will pass the test. Write this as a distance from 80 using the absolute value notation.

Important Features

The most significant feature of the absolute value graph is the corner point where the graph changes direction. When finding the equation for a transformed absolute value function, this point is very helpful for determining the horizontal and vertical shifts.

Example 3

Write an equation for the function graphed.

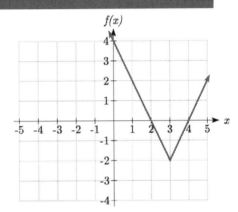

The basic absolute value function changes direction at the origin, so this graph has been shifted to the right 3 and down 2 from the basic toolkit function.

We might also notice that the graph appears stretched, since the linear portions have slopes of 2 and -2. From this information we could write the write the equation in two ways:

$f(x) = 2|x - 3| - 2$, treating the stretch as a vertical stretch

$f(x) = |2(x - 3)| - 2$, treating the stretch as a horizontal compression

Note that these equations are algebraically equivalent – the stretch for an absolute value function can be written interchangeably as a vertical or horizontal stretch/compression.

If you had not been able to determine the stretch based on the slopes of the lines, you can solve for the stretch factor by putting in a known pair of values for x and $f(x)$

$f(x) = a|x - 3| - 2$ Now substituting in the point (1, 2)

$2 = a|1 - 3| - 2$

$4 = 2a$

$a = 2$

Try it Now
2. Given the description of the transformed absolute value function write the equation. The absolute value function is horizontally shifted left 2 units, is vertically flipped, and vertically shifted up 3 units.

The graph of an absolute value function will have a vertical intercept when the input is zero. The graph may or may not have horizontal intercepts, depending on how the graph has been shifted and reflected. It is possible for the absolute value function to have zero, one, or two horizontal intercepts.

Zero horizontal intercepts   One   Two

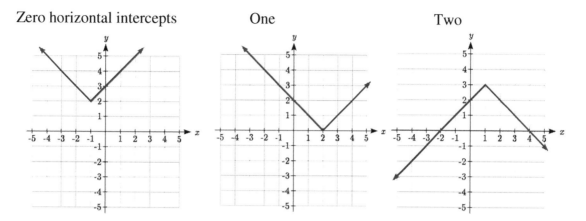

To find the horizontal intercepts, we will need to solve an equation involving an absolute value.

Notice that the absolute value function is not one-to-one, so typically inverses of absolute value functions are not discussed.

Solving Absolute Value Equations

To solve an equation like $8 = |2x - 6|$, we can notice that the absolute value will be equal to eight if the quantity *inside* the absolute value were 8 or -8. This leads to two different equations we can solve independently:

$2x - 6 = 8$ or $2x - 6 = -8$
$2x = 14$ $2x = -2$
$x = 7$ $x = -1$

Solutions to Absolute Value Equations
An equation of the form $|A| = B$, with $B \geq 0$, will have solutions when
$A = B$ or $A = -B$

Example 4

Find the horizontal intercepts of the graph of $f(x) = |4x + 1| - 7$

The horizontal intercepts will occur when $f(x) = 0$. Solving,

$$0 = |4x+1| - 7 \qquad \text{Isolate the absolute value on one side of the equation}$$
$$7 = |4x+1| \qquad \text{Now we can break this into two separate equations:}$$

$$7 = 4x+1 \qquad\qquad -7 = 4x+1$$
$$6 = 4x \qquad \text{or} \qquad -8 = 4x$$
$$x = \frac{6}{4} = \frac{3}{2} \qquad\qquad x = \frac{-8}{4} = -2$$

The graph has two horizontal intercepts, at $x = \dfrac{3}{2}$ and $x = -2$

Example 5

Solve $1 = 4|x-2| + 2$

Isolating the absolute value on one side the equation,
$$1 = 4|x-2| + 2$$
$$-1 = 4|x-2|$$
$$-\frac{1}{4} = |x-2|$$

At this point, we notice that this equation has no solutions – the absolute value always returns a positive value, so it is impossible for the absolute value to equal a negative value.

Try it Now
3. Find the horizontal & vertical intercepts for the function $f(x) = -|x+2| + 3$

Solving Absolute Value Inequalities

When absolute value inequalities are written to describe a set of values, like the inequality $|x-5| \le 4$ we wrote earlier, it is sometimes desirable to express this set of values without the absolute value, either using inequalities, or using interval notation.

We will explore two approaches to solving absolute value inequalities:
1) Using the graph
2) Using test values

Example 6

Solve $|x-5| \le 4$

With both approaches, we will need to know first where the corresponding *equality* is true. In this case, we first will find where $|x-5| = 4$. We do this because the absolute value is a nice friendly function with no breaks, so the only way the function values can switch from being less than 4 to being greater than 4 is by passing through where the values equal 4. Solve $|x-5| = 4$,

$$x-5 = 4 \qquad x-5 = -4$$
$$\qquad\qquad \text{or}$$
$$x = 9 \qquad\qquad x = 1$$

To use a graph, we can sketch the function $f(x) = |x-5|$. To help us see where the outputs are 4, the line $g(x) = 4$ could also be sketched.

On the graph, we can see that indeed the output values of the absolute value are equal to 4 at $x = 1$ and $x = 9$. Based on the shape of the graph, we can determine the absolute value is less than or equal to 4 between these two points, when $1 \le x \le 9$. In interval notation, this would be the interval [1,9].

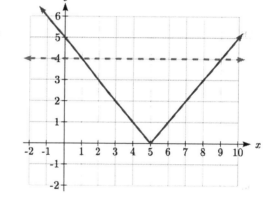

As an alternative to graphing, after determining that the absolute value is equal to 4 at $x = 1$ and $x = 9$, we know the graph can only change from being less than 4 to greater than 4 at these values. This divides the number line up into three intervals: $x<1$, $1<x<9$, and $x>9$. To determine when the function is less than 4, we could pick a value in each interval and see if the output is less than or greater than 4.

Interval	Test x	$f(x)$	<4 or >4?
$x<1$	0	$|0-5| = 5$	greater
$1<x<9$	6	$|6-5| = 1$	less
$x>9$	11	$|11-5| = 6$	greater

Since $1 \le x \le 9$ is the only interval in which the output at the test value is less than 4, we can conclude the solution to $|x-5| \le 4$ is $1 \le x \le 9$.

Example 7

Given the function $f(x) = -\frac{1}{2}|4x-5|+3$, determine for what x values the function values are negative.

We are trying to determine where $f(x) < 0$, which is when $-\frac{1}{2}|4x-5|+3<0$. We begin by isolating the absolute value:

$-\frac{1}{2}|4x-5|<-3$ when we multiply both sides by -2, it reverses the inequality

$|4x-5|>6$

Next we solve for the equality $|4x-5|=6$

$4x-5=6$ $4x-5=-6$

$4x=11$ or $4x=-1$

$x=\dfrac{11}{4}$ $x=\dfrac{-1}{4}$

We can now either pick test values or sketch a graph of the function to determine on which intervals the original function value are negative. Notice that it is not even really important exactly what the graph looks like, as long as we know that it crosses the horizontal axis at $x=\dfrac{-1}{4}$ and $x=\dfrac{11}{4}$, and that the graph has been reflected vertically.

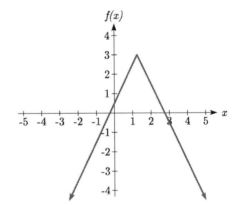

From the graph of the function, we can see the function values are negative to the left of the first horizontal intercept at $x=\dfrac{-1}{4}$, and negative to the right of the second intercept at $x=\dfrac{11}{4}$. This gives us the solution to the inequality:

$x<\dfrac{-1}{4}$ or $x>\dfrac{11}{4}$

In interval notation, this would be $\left(-\infty,\dfrac{-1}{4}\right)\cup\left(\dfrac{11}{4},\infty\right)$

Try it Now

4. Solve $-2|k-4| \le -6$

Important Topics of this Section
The properties of the absolute value function
Solving absolute value equations
Finding intercepts
Solving absolute value inequalities

Try it Now Answers

1. Using the variable p, for passing, $|p - 80| \le 20$

2. $f(x) = -|x+2| + 3$

3. $f(0) = 1$, so the vertical intercept is at (0,1).
   $f(x) = 0$ when
   $-|x+2| + 3 = 0$
   $|x+2| = 3$
   $x+2 = 3$ or $x+2 = -3$
   $x = 1$ or $x = -5$ so the horizontal intercepts are at (-5,0) & (1,0)

4. $-2|k-4| \le -6$

   $|k-4| \ge 3$

   Solving the equality $|k-4| = 3$, $k-4 = 3$ or $k-4 = -3$, so
   $k = 1$ or $k = 7$.
   Using a graph or test values, we can determine the
   intervals that satisfy the inequality are $k \le 1$ or $k \ge 7$; in
   interval notation this would be $(-\infty, 1] \cup [7, \infty)$

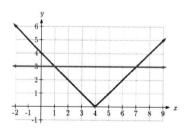

Section 2.5 Exercises

Write an equation for each transformation of $f(x) = |x|$

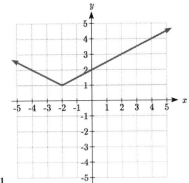

1.

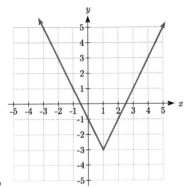

2.

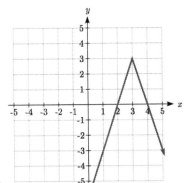

3.

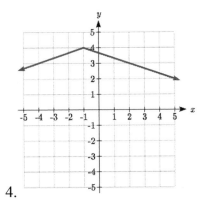

4.

Sketch a graph of each function

5. $f(x) = -|x-1| - 1$

6. $f(x) = -|x+3| + 4$

7. $f(x) = 2|x+3| + 1$

8. $f(x) = 3|x-2| - 3$

9. $f(x) = |2x-4| - 3$

10. $f(x) = |3x+9| + 2$

Solve each the equation

11. $|5x-2| = 11$

12. $|4x+2| = 15$

13. $2|4-x| = 7$

14. $3|5-x| = 5$

15. $3|x+1| - 4 = -2$

16. $5|x-4| - 7 = 2$

Find the horizontal and vertical intercepts of each function

17. $f(x)=2|x+1|-10$ 18. $f(x)=4|x-3|+4$

19. $f(x)=-3|x-2|-1$ 20. $f(x)=-2|x+1|+6$

Solve each inequality

21. $|x+5|<6$ 22. $|x-3|<7$

23. $|x-2|\geq3$ 24. $|x+4|\geq2$

25. $|3x+9|<4$ 26. $|2x-9|\leq8$

Chapter 3: Polynomial and Rational Functions

Section 3.1 Power Functions & Polynomial Functions

A square is cut out of cardboard, with each side having length L. If we wanted to write a function for the area of the square, with L as the input and the area as output, you may recall that the area of a rectangle can be found by multiplying the length times the width. Since our shape is a square, the length & the width are the same, giving the formula:

$$A(L) = L \cdot L = L^2$$

Likewise, if we wanted a function for the volume of a cube with each side having some length L, you may recall volume of a rectangular box can be found by multiplying length by width by height, which are all equal for a cube, giving the formula:

$$V(L) = L \cdot L \cdot L = L^3$$

These two functions are examples of **power functions**, functions that are some power of the variable.

Power Function

A **power function** is a function that can be represented in the form

$$f(x) = x^p$$

Where the base is a variable and the exponent, p, is a number.

Example 1

Which of our toolkit functions are power functions?

The constant and identity functions are power functions, since they can be written as $f(x) = x^0$ and $f(x) = x^1$ respectively.

The quadratic and cubic functions are both power functions with whole number powers: $f(x) = x^2$ and $f(x) = x^3$.

The reciprocal and reciprocal squared functions are both power functions with negative whole number powers since they can be written as $f(x) = x^{-1}$ and $f(x) = x^{-2}$.

The square and cube root functions are both power functions with fractional powers since they can be written as $f(x) = x^{1/2}$ or $f(x) = x^{1/3}$.

Try it Now
1. What point(s) do the toolkit power functions have in common? O, O

Characteristics of Power Functions

Shown to the right are the graphs of
$f(x) = x^2$, $f(x) = x^4$, and $f(x) = x^6$, all even whole number powers. Notice that all these graphs have a fairly similar shape, very similar to the quadratic toolkit, but as the power increases the graphs flatten somewhat near the origin, and become steeper away from the origin.

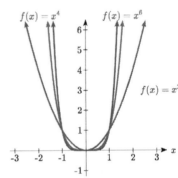

To describe the behavior as numbers become larger and larger, we use the idea of infinity. The symbol for positive infinity is ∞, and $-\infty$ for negative infinity. When we say that "x approaches infinity", which can be symbolically written as $x \to \infty$, we are describing a behavior – we are saying that x is getting large in the positive direction.

With the even power functions, as the x becomes large in either the positive or negative direction, the output values become very large positive numbers. Equivalently, we could describe this by saying that as x approaches positive or negative infinity, the f(x) values approach positive infinity. In symbolic form, we could write: as $x \to \pm\infty$, $f(x) \to \infty$.

Shown here are the graphs of
$f(x) = x^3$, $f(x) = x^5$, and $f(x) = x^7$, all odd whole number powers. Notice all these graphs look similar to the cubic toolkit, but again as the power increases the graphs flatten near the origin and become steeper away from the origin.

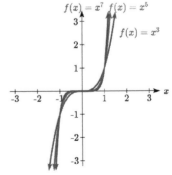

For these odd power functions, as x approaches negative infinity, f(x) approaches negative infinity. As x approaches positive infinity, f(x) approaches positive infinity. In symbolic form we write: as $x \to -\infty$, $f(x) \to -\infty$ and as $x \to \infty$, $f(x) \to \infty$.

Long Run Behavior

The behavior of the graph of a function as the input takes on large negative values, $x \to -\infty$, and large positive values, $x \to \infty$, is referred to as the **long run behavior** of the function.

Example 2

Describe the long run behavior of the graph of $f(x) = x^8$.

Since $f(x) = x^8$ has a whole, even power, we would expect this function to behave somewhat like the quadratic function. As the input gets large positive or negative, we would expect the output to grow without bound in the positive direction. In symbolic form, as $x \to \pm\infty$, $f(x) \to \infty$.

Example 3

Describe the long run behavior of the graph of $f(x) = -x^9$

Since this function has a whole odd power, we would expect it to behave somewhat like the cubic function. The negative in front of the x^9 will cause a vertical reflection, so as the inputs grow large positive, the outputs will grow large in the negative direction, and as the inputs grow large negative, the outputs will grow large in the positive direction. In symbolic form, for the long run behavior we would write: as $x \to \infty$, $f(x) \to -\infty$ and as $x \to -\infty$, $f(x) \to \infty$.

You may use words or symbols to describe the long run behavior of these functions.

Try it Now
2. Describe in words and symbols the long run behavior of $f(x) = -x^4$

Treatment of the rational and radical forms of power functions will be saved for later.

Polynomials

An oil pipeline bursts in the Gulf of Mexico, causing an oil slick in a roughly circular shape. The slick is currently 24 miles in radius, but that radius is increasing by 8 miles each week. If we wanted to write a formula for the area covered by the oil slick, we could do so by composing two functions together. The first is a formula for the radius, r, of the spill, which depends on the number of weeks, w, that have passed.

Hopefully you recognized that this relationship is linear:
$$r(w) = 24 + 8w$$

We can combine this with the formula for the area, A, of a circle:
$$A(r) = \pi r^2$$

Composing these functions gives a formula for the area in terms of weeks:
$$A(w) = A(r(w)) = A(24 + 8w) = \pi(24 + 8w)^2$$

Multiplying this out gives the formula
$$A(w) = 576\pi + 384\pi w + 64\pi w^2$$

This formula is an example of a **polynomial**. A polynomial is simply the sum of terms each consisting of a transformed power function with positive whole number power.

Terminology of Polynomial Functions

A **polynomial** is function that can be written as $f(x) = a_0 + a_1 x + a_2 x^2 + \cdots + a_n x^n$

Each of the a_i constants are called **coefficients** and can be positive, negative, or zero, and be whole numbers, decimals, or fractions.

A **term** of the polynomial is any one piece of the sum, that is any $a_i x^i$. Each individual term is a transformed power function.

The **degree** of the polynomial is the highest power of the variable that occurs in the polynomial.

The **leading term** is the term containing the highest power of the variable: the term with the highest degree.

The **leading coefficient** is the coefficient of the leading term.

Because of the definition of the "leading" term we often rearrange polynomials so that the powers are descending.

$f(x) = a_n x^n + \ldots + a_2 x^2 + a_1 x + a_0$

Example 4

Identify the degree, leading term, and leading coefficient of these polynomials:

a) $f(x) = 3 + 2x^2 - 4x^3$ b) $g(t) = 5t^5 - 2t^3 + 7t$ c) $h(p) = 6p - p^3 - 2$

a) For the function $f(x)$, the degree is 3, the highest power on x. The leading term is the term containing that power, $-4x^3$. The leading coefficient is the coefficient of that term, -4.

b) For $g(t)$, the degree is 5, the leading term is $5t^5$, and the leading coefficient is 5.

c) For $h(p)$, the degree is 3, the leading term is $-p^3$, so the leading coefficient is -1.

Long Run Behavior of Polynomials

For any polynomial, the **long run behavior** of the polynomial will match the long run behavior of the leading term.

Example 5

What can we determine about the long run behavior and degree of the equation for the polynomial graphed here?

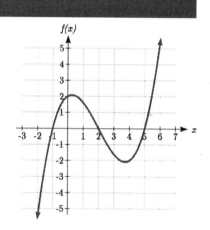

Since the output grows large and positive as the inputs grow large and positive, we describe the long run behavior symbolically by writing: as $x \to \infty$, $f(x) \to \infty$. Similarly, as $x \to -\infty$, $f(x) \to -\infty$.

In words, we could say that as x values approach infinity, the function values approach infinity, and as x values approach negative infinity the function values approach negative infinity.

We can tell this graph has the shape of an odd degree power function which has not been reflected, so the degree of the polynomial creating this graph must be odd, and the leading coefficient would be positive.

Try it Now

3. Given the function $f(x) = 0.2(x-2)(x+1)(x-5)$ use your algebra skills to write the function in standard polynomial form (as a sum of terms) and determine the leading term, degree, and long run behavior of the function.

Short Run Behavior

Characteristics of the graph such as vertical and horizontal intercepts and the places the graph changes direction are part of the short run behavior of the polynomial.

Like with all functions, the vertical intercept is where the graph crosses the vertical axis, and occurs when the input value is zero. Since a polynomial is a function, there can only be one vertical intercept, which occurs at the point $(0, a_0)$. The horizontal intercepts occur at the input values that correspond with an output value of zero. It is possible to have more than one horizontal intercept.

Horizontal intercepts are also called **zeros**, or **roots** of the function.

Example 6

Given the polynomial function $f(x) = (x - 2)(x + 1)(x - 4)$, written in factored form for your convenience, determine the vertical and horizontal intercepts.

The vertical intercept occurs when the input is zero.
$f(0) = (0 - 2)(0 + 1)(0 - 4) = 8$.

The graph crosses the vertical axis at the point (0, 8).

The horizontal intercepts occur when the output is zero.
$0 = (x - 2)(x + 1)(x - 4)$ when $x = 2$, -1, or 4.
f(x) has zeros, or roots, at $x = 2$, -1, and 4.

The graph crosses the horizontal axis at the points (2, 0), (-1, 0), and (4, 0)

Notice that the polynomial in the previous example, which would be degree three if multiplied out, had three horizontal intercepts and two turning points – places where the graph changes direction. We will now make a general statement without justifying it – the reasons will become clear later in this chapter.

Intercepts and Turning Points of Polynomials

A polynomial of degree *n* will have:

At most *n* horizontal intercepts. An odd degree polynomial will always have at least one.

At most *n*−1 turning points

Example 7

What can we conclude about the graph of the polynomial shown here?

Based on the long run behavior, with the graph becoming large positive on both ends of the graph, we can determine that this is the graph of an even degree polynomial. The graph has 2 horizontal intercepts, suggesting a degree of 2 or greater, and 3 turning points, suggesting a degree of 4 or greater. Based on this, it would be reasonable to conclude that the degree is even and at least 4, so it is probably a fourth degree polynomial.

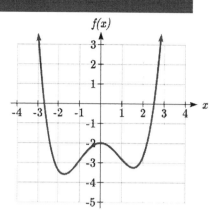

Try it Now

4. Given the function $f(x) = 0.2(x-2)(x+1)(x-5)$, determine the short run behavior.

Important Topics of this Section
Power Functions
Polynomials
Coefficients
Leading coefficient
Term
Leading Term
Degree of a polynomial
Long run behavior
Short run behavior

Try it Now Answers

1. (0, 0) and (1, 1) are common to all power functions.
2. As x approaches positive and negative infinity, $f(x)$ approaches negative infinity: as $x \to \pm\infty$, $f(x) \to -\infty$ because of the vertical flip.
3. The leading term is $0.2x^3$, so it is a degree 3 polynomial.
    As x approaches infinity (or gets very large in the positive direction) $f(x)$ approaches infinity; as x approaches negative infinity (or gets very large in the negative direction) $f(x)$ approaches negative infinity. (Basically the long run behavior is the same as the cubic function).
4. Horizontal intercepts are (2, 0) (-1, 0) and (5, 0), the vertical intercept is (0, 2) and there are 2 turns in the graph.

Section 3.1 Exercises

Find the long run behavior of each function as $x \to \infty$ and $x \to -\infty$

1. $f(x) = x^4$ 2. $f(x) = x^6$ 3. $f(x) = x^3$ 4. $f(x) = x^5$

5. $f(x) = -x^2$ 6. $f(x) = -x^4$ 7. $f(x) = -x^7$ 8. $f(x) = -x^9$

Find the degree and leading coefficient of each polynomial

9. $4x^7$ 10. $5x^6$

11. $5 - x^2$ 12. $6 + 3x - 4x^3$

13. $-2x^4 - 3x^2 + x - 1$ 14. $6x^5 - 2x^4 + x^2 + 3$

15. $(2x+3)(x-4)(3x+1)$ 16. $(3x+1)(x+1)(4x+3)$

Find the long run behavior of each function as $x \to \infty$ and $x \to -\infty$

17. $-2x^4 - 3x^2 + x - 1$ 18. $6x^5 - 2x^4 + x^2 + 3$

19. $3x^2 + x - 2$ 20. $-2x^3 + x^2 - x + 3$

21. What is the maximum number of x-intercepts and turning points for a polynomial of degree 5?

22. What is the maximum number of x-intercepts and turning points for a polynomial of degree 8?

What is the least possible degree of the polynomial function shown in each graph?

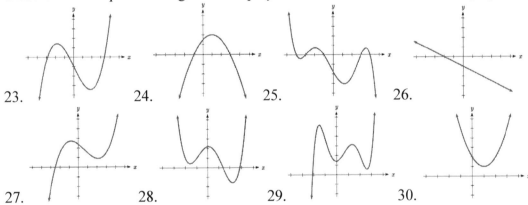

23.   24.   25.   26.

27.   28.   29.   30.

Find the vertical and horizontal intercepts of each function.

31. $f(t) = 2(t-1)(t+2)(t-3)$ 32. $f(x) = 3(x+1)(x-4)(x+5)$

33. $g(n) = -2(3n-1)(2n+1)$ 34. $k(u) = -3(4-n)(4n+3)$

Section 3.2 Quadratic Functions

In this section, we will explore the family of 2nd degree polynomials, the quadratic functions. While they share many characteristics of polynomials in general, the calculations involved in working with quadratics is typically a little simpler, which makes them a good place to start our exploration of short run behavior. In addition, quadratics commonly arise from problems involving area and projectile motion, providing some interesting applications.

Example 1

A backyard farmer wants to enclose a rectangular space for a new garden. She has purchased 80 feet of wire fencing to enclose 3 sides, and will put the 4th side against the backyard fence. Find a formula for the area enclosed by the fence if the sides of fencing perpendicular to the existing fence have length L.

In a scenario like this involving geometry, it is often helpful to draw a picture. It might also be helpful to introduce a temporary variable, W, to represent the side of fencing parallel to the 4th side or backyard fence.

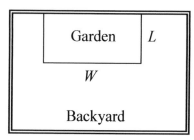

Since we know we only have 80 feet of fence available, we know that $L + W + L = 80$, or more simply, $2L + W = 80$. This allows us to represent the width, W, in terms of L: $W = 80 - 2L$

Now we are ready to write an equation for the area the fence encloses. We know the area of a rectangle is length multiplied by width, so
$A = LW = L(80 - 2L)$
$A(L) = 80L - 2L^2$
This formula represents the area of the fence in terms of the variable length L.

Short run Behavior: Vertex

We now explore the interesting features of the graphs of quadratics. In addition to intercepts, quadratics have an interesting feature where they change direction, called the **vertex**. You probably noticed that all quadratics are related to transformations of the basic quadratic function $f(x) = x^2$.

Example 2

Write an equation for the quadratic graphed below as a transformation of $f(x) = x^2$, then expand the formula and simplify terms to write the equation in standard polynomial form.

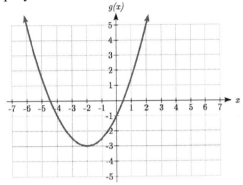

We can see the graph is the basic quadratic shifted to the left 2 and down 3, giving a formula in the form $g(x) = a(x+2)^2 - 3$. By plugging in a point that falls on the grid, such as $(0,-1)$, we can solve for the stretch factor:

$-1 = a(0+2)^2 - 3$

$2 = 4a$

$a = \dfrac{1}{2}$

Written as a transformation, the equation for this formula is $g(x) = \dfrac{1}{2}(x+2)^2 - 3$. To write this in standard polynomial form, we can expand the formula and simplify terms:

$g(x) = \dfrac{1}{2}(x+2)^2 - 3$

$g(x) = \dfrac{1}{2}(x+2)(x+2) - 3$

$g(x) = \dfrac{1}{2}(x^2 + 4x + 4) - 3$

$g(x) = \dfrac{1}{2}x^2 + 2x + 2 - 3$

$g(x) = \dfrac{1}{2}x^2 + 2x - 1$

Notice that the horizontal and vertical shifts of the basic quadratic determine the location of the vertex of the parabola; the vertex is unaffected by stretches and compressions.

Try it Now
1.   A coordinate grid has been superimposed over the quadratic path of a basketball[1]. Find an equation for the path of the ball. Does he make the basket?

Forms of Quadratic Functions

The **standard form** of a quadratic function is $f(x) = ax^2 + bx + c$

The **transformation form** of a quadratic function is $f(x) = a(x - h)^2 + k$

The **vertex** of the quadratic function is located at (h, k), where h and k are the numbers in the transformation form of the function. Because the vertex appears in the transformation form, it is often called the **vertex form**.

In the previous example, we saw that it is possible to rewrite a quadratic function given in transformation form and rewrite it in standard form by expanding the formula. It would be useful to reverse this process, since the transformation form reveals the vertex.

Expanding out the general transformation form of a quadratic gives:
$$f(x) = a(x - h)^2 + k = a(x - h)(x - h) + k$$
$$f(x) = a(x^2 - 2xh + h^2) + k = ax^2 - 2ahx + ah^2 + k$$

This should be equal to the standard form of the quadratic:
$$ax^2 - 2ahx + ah^2 + k = ax^2 + bx + c$$

The second degree terms are already equal. For the linear terms to be equal, the coefficients must be equal:
$$-2ah = b, \text{ so } h = -\frac{b}{2a}$$

This provides us a method to determine the horizontal shift of the quadratic from the standard form. We could likewise set the constant terms equal to find:
$$ah^2 + k = c, \text{ so } k = c - ah^2 = c - a\left(-\frac{b}{2a}\right)^2 = c - a\frac{b^2}{4a^2} = c - \frac{b^2}{4a}$$

In practice, though, it is usually easier to remember that k is the output value of the function when the input is h, so $k = f(h)$.

Finding the Vertex of a Quadratic

For a quadratic given in standard form, the vertex (h, k) is located at:

$$h = -\frac{b}{2a}, \quad k = f(h) = f\left(\frac{-b}{2a}\right)$$

Example 3

Find the vertex of the quadratic $f(x) = 2x^2 - 6x + 7$. Rewrite the quadratic into transformation form (vertex form).

The horizontal coordinate of the vertex will be at $h = -\dfrac{b}{2a} = -\dfrac{-6}{2(2)} = \dfrac{6}{4} = \dfrac{3}{2}$

The vertical coordinate of the vertex will be at $f\left(\dfrac{3}{2}\right) = 2\left(\dfrac{3}{2}\right)^2 - 6\left(\dfrac{3}{2}\right) + 7 = \dfrac{5}{2}$

Rewriting into transformation form, the stretch factor will be the same as the a in the original quadratic. Using the vertex to determine the shifts,

$$f(x) = 2\left(x - \frac{3}{2}\right)^2 + \frac{5}{2}$$

Try it Now

2. Given the equation $g(x) = 13 + x^2 - 6x$ write the equation in standard form and then in transformation/vertex form.

As an alternative to using a formula for finding the vertex, the equation can also be written into vertex form by **completing the square**. This process is most easily explained through example. In most cases, using the formula for finding the vertex will be quicker and easier than completing the square, but completing the square is a useful technique when faced with some other algebraic problems.

Example 4

Rewrite $f(x) = 2x^2 - 12x + 14$ into vertex form by completing the square.

We start by factoring the leading coefficient from the quadratic and linear terms.

$2\left(x^2 - 6x\right) + 14$

Next, we are going to add something inside the parentheses so that the quadratic inside the parentheses becomes a perfect square. In other words, we are looking for values p and q so that $\left(x^2 - 6x + p\right) = (x - q)^2$.

Notice that if multiplied out on the right, the middle term would be $-2q$, so q must be half of the middle term on the left; $q = -3$. In that case, p must be $(-3)^2 = 9$.
$$\left(x^2 - 6x + 9\right) = (x - 3)^2$$

Now, we can't just add 9 into the expression – that would change the value of the expression. In fact, adding 9 inside the parentheses actually adds 18 to the expression, since the 2 outside the parentheses will distribute. To keep the expression balanced, we can subtract 18.
$$2\left(x^2 - 6x + 9\right) + 14 - 18$$

Simplifying, we are left with vertex form.
$$2(x - 3)^2 - 4$$

In addition to enabling us to more easily graph a quadratic written in standard form, finding the vertex serves another important purpose – it allows us to determine the maximum or minimum value of the function, depending on which way the graph opens.

Example 5

Returning to our backyard farmer from the beginning of the section, what dimensions should she make her garden to maximize the enclosed area?

Earlier we determined the area she could enclose with 80 feet of fencing on three sides was given by the equation $A(L) = 80L - 2L^2$. Notice that quadratic has been vertically reflected, since the coefficient on the squared term is negative, so the graph will open downwards, and the vertex will be a maximum value for the area.

In finding the vertex, we take care since the equation is not written in standard polynomial form with decreasing powers. But we know that a is the coefficient on the squared term, so $a = -2$, $b = 80$, and $c = 0$.
Finding the vertex:
$$h = -\frac{80}{2(-2)} = 20, \quad k = A(20) = 80(20) - 2(20)^2 = 800$$

The maximum value of the function is an area of 800 square feet, which occurs when L = 20 feet. When the shorter sides are 20 feet, that leaves 40 feet of fencing for the longer side. To maximize the area, she should enclose the garden so the two shorter sides have length 20 feet, and the longer side parallel to the existing fence has length 40 feet.

Example 6

A local newspaper currently has 84,000 subscribers, at a quarterly charge of $30. Market research has suggested that if they raised the price to $32, they would lose 5,000 subscribers. Assuming that subscriptions are linearly related to the price, what price should the newspaper charge for a quarterly subscription to maximize their revenue?

Revenue is the amount of money a company brings in. In this case, the revenue can be found by multiplying the charge per subscription times the number of subscribers. We can introduce variables, C for charge per subscription and S for the number subscribers, giving us the equation
Revenue $= CS$

Since the number of subscribers changes with the price, we need to find a relationship between the variables. We know that currently $S = 84,000$ and $C = 30$, and that if they raise the price to $32 they would lose 5,000 subscribers, giving a second pair of values, $C = 32$ and $S = 79,000$. From this we can find a linear equation relating the two quantities. Treating C as the input and S as the output, the equation will have form $S = mC + b$. The slope will be
$$m = \frac{79,000 - 84,000}{32 - 30} = \frac{-5,000}{2} = -2,500$$

This tells us the paper will lose 2,500 subscribers for each dollar they raise the price. We can then solve for the vertical intercept

$S = -2500C + b$	Plug in the point $S = 84,000$ and $C = 30$
$84,000 = -2500(30) + b$	Solve for b
$b = 159,000$	

This gives us the linear equation $S = -2,500C + 159,000$ relating cost and subscribers. We now return to our revenue equation.

Revenue $= CS$	Substituting the equation for S from above
Revenue $= C(-2,500C + 159,000)$	Expanding
Revenue $= -2,500C^2 + 159,000C$	

We now have a quadratic equation for revenue as a function of the subscription charge. To find the price that will maximize revenue for the newspaper, we can find the vertex:
$$h = -\frac{159,000}{2(-2,500)} = 31.8$$

The model tells us that the maximum revenue will occur if the newspaper charges $31.80 for a subscription. To find what the maximum revenue is, we can evaluate the revenue equation:
Maximum Revenue $= -2,500(31.8)^2 + 159,000(31.8) = \$2,528,100$

Short run Behavior: Intercepts

As with any function, we can find the vertical intercepts of a quadratic by evaluating the function at an input of zero, and we can find the horizontal intercepts by solving for when the output will be zero. Notice that depending upon the location of the graph, we might have zero, one, or two horizontal intercepts.

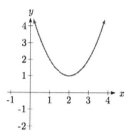
zero horizontal intercepts

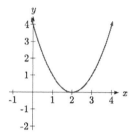
one horizontal intercept

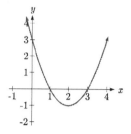
two horizontal intercepts

Example 7

Find the vertical and horizontal intercepts of the quadratic $f(x) = 3x^2 + 5x - 2$

We can find the vertical intercept by evaluating the function at an input of zero:
$f(0) = 3(0)^2 + 5(0) - 2 = -2$ Vertical intercept at (0,-2)

For the horizontal intercepts, we solve for when the output will be zero
$0 = 3x^2 + 5x - 2$

In this case, the quadratic can be factored easily, providing the simplest method for solution
$0 = (3x - 1)(x + 2)$
$0 = 3x - 1$
$x = \dfrac{1}{3}$ or $\begin{aligned}0 &= x + 2 \\ x &= -2\end{aligned}$ Horizontal intercepts at $\left(\dfrac{1}{3}, 0\right)$ and (-2,0)

Notice that in the standard form of a quadratic, the constant term c reveals the vertical intercept of the graph.

Example 8

Find the horizontal intercepts of the quadratic $f(x) = 2x^2 + 4x - 4$

Again we will solve for when the output will be zero
$0 = 2x^2 + 4x - 4$

Since the quadratic is not easily factorable in this case, we solve for the intercepts by first rewriting the quadratic into transformation form.

$$h = -\frac{b}{2a} = -\frac{4}{2(2)} = -1 \qquad k = f(-1) = 2(-1)^2 + 4(-1) - 4 = -6$$

$$f(x) = 2(x+1)^2 - 6$$

Now we can solve for when the output will be zero

$$0 = 2(x+1)^2 - 6$$

$$6 = 2(x+1)^2$$

$$3 = (x+1)^2$$

$$x + 1 = \pm\sqrt{3}$$

$$x = -1 \pm \sqrt{3}$$

The graph has horizontal intercepts at $(-1 - \sqrt{3}, 0)$ and $(-1 + \sqrt{3}, 0)$

Try it Now

3. In Try it Now problem 2 we found the standard & transformation form for the function $g(x) = 13 + x^2 - 6x$. Now find the Vertical & Horizontal intercepts (if any).

The process in the last example is done commonly enough that sometimes people find it easier to solve the problem once in general and remember the formula for the result, rather than repeating the process each time. Based on our previous work we showed that any quadratic in standard form can be written into transformation form as:

$$f(x) = a\left(x + \frac{b}{2a}\right)^2 + c - \frac{b^2}{4a}$$

Solving for the horizontal intercepts using this general equation gives:

$$0 = a\left(x + \frac{b}{2a}\right)^2 + c - \frac{b^2}{4a} \qquad \text{start to solve for } x \text{ by moving the constants to the other side}$$

$$\frac{b^2}{4a} - c = a\left(x + \frac{b}{2a}\right)^2 \qquad \text{divide both sides by } a$$

$$\frac{b^2}{4a^2} - \frac{c}{a} = \left(x + \frac{b}{2a}\right)^2 \qquad \text{find a common denominator to combine fractions}$$

$$\frac{b^2}{4a^2} - \frac{4ac}{4a^2} = \left(x + \frac{b}{2a}\right)^2 \qquad \text{combine the fractions on the left side of the equation}$$

$$\frac{b^2 - 4ac}{4a^2} = \left(x + \frac{b}{2a}\right)^2$$ take the square root of both sides

$$\pm\sqrt{\frac{b^2 - 4ac}{4a^2}} = x + \frac{b}{2a}$$ subtract $b/2a$ from both sides

$$-\frac{b}{2a} \pm \frac{\sqrt{b^2 - 4ac}}{2a} = x$$ combining the fractions

$$x = \frac{-b \pm \sqrt{b^2 - 4ac}}{2a}$$ Notice that this can yield two different answers for x

---

Quadratic Formula

For a quadratic function given in standard form $f(x) = ax^2 + bx + c$, the **quadratic formula** gives the horizontal intercepts of the graph of this function.

$$x = \frac{-b \pm \sqrt{b^2 - 4ac}}{2a}$$

Example 9

A ball is thrown upwards from the top of a 40-foot-tall building at a speed of 80 feet per second. The ball's height above ground can be modeled by the equation
$H(t) = -16t^2 + 80t + 40$.
What is the maximum height of the ball?
When does the ball hit the ground?

To find the maximum height of the ball, we would need to know the vertex of the quadratic.

$$h = -\frac{80}{2(-16)} = \frac{80}{32} = \frac{5}{2}, \quad k = H\left(\frac{5}{2}\right) = -16\left(\frac{5}{2}\right)^2 + 80\left(\frac{5}{2}\right) + 40 = 140$$

The ball reaches a maximum height of 140 feet after 2.5 seconds.
To find when the ball hits the ground, we need to determine when the height is zero –
when $H(t) = 0$. While we could do this using the transformation form of the quadratic, we can also use the quadratic formula:

$$t = \frac{-80 \pm \sqrt{80^2 - 4(-16)(40)}}{2(-16)} = \frac{-80 \pm \sqrt{8960}}{-32}$$

Since the square root does not simplify nicely, we can use a calculator to approximate the values of the solutions:

$$t = \frac{-80 - \sqrt{8960}}{-32} \approx 5.458 \quad \text{or} \quad t = \frac{-80 + \sqrt{8960}}{-32} \approx -0.458$$

The second answer is outside the reasonable domain of our model, so we conclude the ball will hit the ground after about 5.458 seconds.

Try it Now

4. For these two equations determine if the vertex will be a maximum value or a minimum value.

   a. $g(x) = -8x + x^2 + 7$

   b. $g(x) = -3(3 - x)^2 + 2$

Important Topics of this Section

Quadratic functions

    Standard form

    Transformation form/Vertex form

    Vertex as a maximum / Vertex as a minimum

Short run behavior

    Vertex / Horizontal & Vertical intercepts

Quadratic formula

Try it Now Answers

1. The path passes through the origin with vertex at (-4, 7).

$h(x) = -\frac{7}{16}(x + 4)^2 + 7$. To make the shot, $h(-7.5)$ would need to be about 4. $h(-7.5) \approx 1.64$; he doesn't make it.

2. $g(x) = x^2 - 6x + 13$ in Standard form;

Finding the vertex, $h = \frac{-(-6)}{2(1)} = 3$. $k = g(3) = 3^2 - 6(3) + 13 = 4$.

$g(x) = (x - 3)^2 + 4$ in Transformation form

3. Vertical intercept at (0, 13), No horizontal intercepts since the vertex is above the x-axis and the graph opens upwards.

4. a. Vertex is a minimum value, since $a > 0$ and the graph opens upwards
   b. Vertex is a maximum value, since $a < 0$ and the graph opens downwards

Section 3.2 Exercises

Write an equation for the quadratic function graphed.

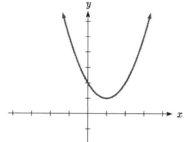

1.

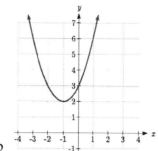

2.

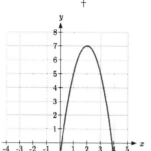

3.

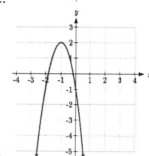

4.

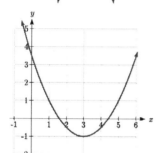

5.

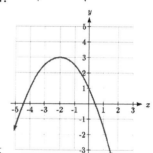

6.

For each of the follow quadratic functions, find a) the vertex, b) the vertical intercept, and c) the horizontal intercepts.

7. $y(x) = 2x^2 + 10x + 12$

8. $z(p) = 3x^2 + 6x - 9$

9. $f(x) = 2x^2 - 10x + 4$

10. $g(x) = -2x^2 - 14x + 12$

11. $h(t) = -4t^2 + 6t - 1$

12. $k(t) = 2x^2 + 4x - 15$

Rewrite the quadratic function into vertex form.

13. $f(x) = x^2 - 12x + 32$

14. $g(x) = x^2 + 2x - 3$

15. $h(x) = 2x^2 + 8x - 10$

16. $k(x) = 3x^2 - 6x - 9$

17. Find the values of b and c so $f(x) = -8x^2 + bx + c$ has vertex $(2, -7)$

18. Find the values of b and c so $f(x) = 6x^2 + bx + c$ has vertex $(7, -9)$

Write an equation for a quadratic with the given features

19. x-intercepts (-3, 0) and (1, 0), and y intercept (0, 2)
20. x-intercepts (2, 0) and (-5, 0), and y intercept (0, 3)
21. x-intercepts (2, 0) and (5, 0), and y intercept (0, 6)
22. x-intercepts (1, 0) and (3, 0), and y intercept (0, 4)
23. Vertex at (4, 0), and y intercept (0, -4)
24. Vertex at (5, 6), and y intercept (0, -1)
25. Vertex at (-3, 2), and passing through (3, -2)
26. Vertex at (1, -3), and passing through (-2, 3)

27. A rocket is launched in the air. Its height, in meters above sea level, as a function of time, in seconds, is given by $h(t) = -4.9t^2 + 229t + 234$.
     a. From what height was the rocket launched?
     b. How high above sea level does the rocket reach its peak?
     c. Assuming the rocket will splash down in the ocean, at what time does splashdown occur?

28. A ball is thrown in the air from the top of a building. Its height, in meters above ground, as a function of time, in seconds, is given by $h(t) = -4.9t^2 + 24t + 8$.
     a. From what height was the ball thrown?
     b. How high above ground does the ball reach its peak?
     c. When does the ball hit the ground?

29. The height of a ball thrown in the air is given by $h(x) = -\dfrac{1}{12}x^2 + 6x + 3$, where x is the horizontal distance in feet from the point at which the ball is thrown.
     a. How high is the ball when it was thrown?
     b. What is the maximum height of the ball?
     c. How far from the thrower does the ball strike the ground?

30. A javelin is thrown in the air. Its height is given by $h(x) = -\dfrac{1}{20}x^2 + 8x + 6$, where x is the horizontal distance in feet from the point at which the javelin is thrown.
     a. How high is the javelin when it was thrown?
     b. What is the maximum height of the javelin?
     c. How far from the thrower does the javelin strike the ground?

31. A box with a square base and no top is to be made from a square piece of cardboard by cutting 6 in. squares out of each corner and folding up the sides. The box needs to hold 1000 in³. How big a piece of cardboard is needed?

32. A box with a square base and no top is to be made from a square piece of cardboard by cutting 4 in. squares out of each corner and folding up the sides. The box needs to hold 2700 in³. How big a piece of cardboard is needed?

33. A farmer wishes to enclose two pens with fencing, as shown. If the farmer has 500 feet of fencing to work with, what dimensions will maximize the area enclosed?

34. A farmer wishes to enclose three pens with fencing, as shown. If the farmer has 700 feet of fencing to work with, what dimensions will maximize the area enclosed?

35. You have a wire that is 56 cm long. You wish to cut it into two pieces. One piece will be bent into the shape of a square. The other piece will be bent into the shape of a circle. Let A represent the total area enclosed by the square and the circle. What is the circumference of the circle when A is a minimum?

36. You have a wire that is 71 cm long. You wish to cut it into two pieces. One piece will be bent into the shape of a right triangle with legs of equal length. The other piece will be bent into the shape of a circle. Let A represent the total area enclosed by the triangle and the circle. What is the circumference of the circle when A is a minimum?

37. A soccer stadium holds 62,000 spectators. With a ticket price of $11, the average attendance has been 26,000. When the price dropped to $9, the average attendance rose to 31,000. Assuming that attendance is linearly related to ticket price, what ticket price would maximize revenue?

38. A farmer finds that if she plants 75 trees per acre, each tree will yield 20 bushels of fruit. She estimates that for each additional tree planted per acre, the yield of each tree will decrease by 3 bushels. How many trees should she plant per acre to maximize her harvest?

39. A hot air balloon takes off from the edge of a mountain lake. Impose a coordinate system as pictured and assume that the path of the balloon follows the graph of

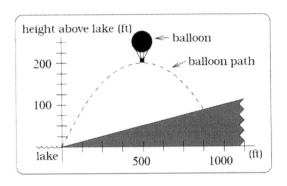

$f(x) = -\dfrac{2}{2500}x^2 + \dfrac{4}{5}x$. The land rises at a constant incline from the lake at the rate of 2 vertical feet for each 20 horizontal feet. [UW]

 a. What is the maximum height of the balloon above water level?
 b. What is the maximum height of the balloon above ground level?
 c. Where does the balloon land on the ground?
 d. Where is the balloon 50 feet above the ground?

40. A hot air balloon takes off from the edge of a plateau. Impose a coordinate system as pictured below and assume that the path the balloon follows is the graph of the quadratic function

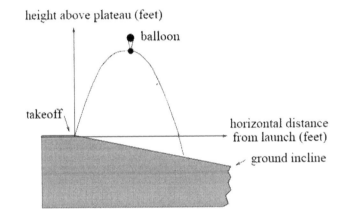

$f(x) = -\dfrac{4}{2500}x^2 + \dfrac{4}{5}x$. The land drops at a constant incline from the plateau at the rate of 1 vertical foot for each 5 horizontal feet. [UW]

 a. What is the maximum height of the balloon above plateau level?
 b. What is the maximum height of the balloon above ground level?
 c. Where does the balloon land on the ground?
 d. Where is the balloon 50 feet above the ground?

Section 3.3 Graphs of Polynomial Functions

In the previous section, we explored the short run behavior of quadratics, a special case of polynomials. In this section, we will explore the short run behavior of polynomials in general.

Short run Behavior: Intercepts

As with any function, the vertical intercept can be found by evaluating the function at an input of zero. Since this is evaluation, it is relatively easy to do it for a polynomial of any degree.

To find horizontal intercepts, we need to solve for when the output will be zero. For general polynomials, this can be a challenging prospect. While quadratics can be solved using the relatively simple quadratic formula, the corresponding formulas for cubic and 4th degree polynomials are not simple enough to remember, and formulas do not exist for general higher-degree polynomials. Consequently, we will limit ourselves to three cases:

1) The polynomial can be factored using known methods: greatest common factor and trinomial factoring.
2) The polynomial is given in factored form.
3) Technology is used to determine the intercepts.

Other techniques for finding the intercepts of general polynomials will be explored in the next section.

Example 1

Find the horizontal intercepts of $f(x) = x^6 - 3x^4 + 2x^2$.

We can attempt to factor this polynomial to find solutions for $f(x) = 0$.

$x^6 - 3x^4 + 2x^2 = 0$ Factoring out the greatest common factor

$x^2(x^4 - 3x^2 + 2) = 0$ Factoring the inside as a quadratic in x^2

$x^2(x^2 - 1)(x^2 - 2) = 0$ Then break apart to find solutions

$$(x^2 - 1) = 0 \qquad\qquad (x^2 - 2) = 0$$

$x^2 = 0$ or $x^2 = 1$ or $x^2 = 2$

$x = 0$ $x = \pm 1$ $x = \pm\sqrt{2}$

This gives us 5 horizontal intercepts.

Example 2

Find the vertical and horizontal intercepts of $g(t) = (t-2)^2(2t+3)$

The vertical intercept can be found by evaluating $g(0)$.
$g(0) = (0-2)^2(2(0)+3) = 12$

The horizontal intercepts can be found by solving $g(t) = 0$
$(t-2)^2(2t+3) = 0$ Since this is already factored, we can break it apart:

$(t-2)^2 = 0$ $(2t+3) = 0$

$t-2 = 0$ or $t = \dfrac{-3}{2}$

$t = 2$

We can always check our answers are reasonable by graphing the polynomial.

Example 3

Find the horizontal intercepts of $h(t) = t^3 + 4t^2 + t - 6$

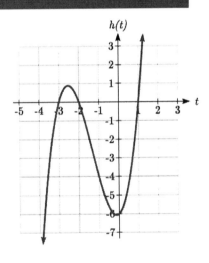

Since this polynomial is not in factored form, has no common factors, and does not appear to be factorable using techniques we know, we can turn to technology to find the intercepts.

Graphing this function, it appears there are horizontal intercepts at $t = -3, -2,$ and 1.

We could check these are correct by plugging in these values for t and verifying that $h(-3) = h(-2) = h(1) = 0$.

Try it Now

1. Find the vertical and horizontal intercepts of the function $f(t) = t^4 - 4t^2$.

Graphical Behavior at Intercepts

If we graph the function $f(x) = (x+3)(x-2)^2(x+1)^3$, notice that the behavior at each of the horizontal intercepts is different.

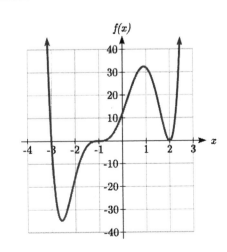

At the horizontal intercept $x = -3$, coming from the $(x+3)$ factor of the polynomial, the graph passes directly through the horizontal intercept.

The factor $(x+3)$ is linear (has a power of 1), so the behavior near the intercept is like that of a line - it passes directly through the intercept. We call this a single zero, since the zero corresponds to a single factor of the function.

At the horizontal intercept $x = 2$, coming from the $(x-2)^2$ factor of the polynomial, the graph touches the axis at the intercept and changes direction. The factor is quadratic (degree 2), so the behavior near the intercept is like that of a quadratic – it bounces off the horizontal axis at the intercept. Since $(x-2)^2 = (x-2)(x-2)$, the factor is repeated twice, so we call this a double zero. We could also say the zero has **multiplicity** 2.

At the horizontal intercept $x = -1$, coming from the $(x+1)^3$ factor of the polynomial, the graph passes through the axis at the intercept, but flattens out a bit first. This factor is cubic (degree 3), so the behavior near the intercept is like that of a cubic, with the same "S" type shape near the intercept that the toolkit x^3 has. We call this a triple zero. We could also say the zero has multiplicity 3.

By utilizing these behaviors, we can sketch a reasonable graph of a factored polynomial function without needing technology.

Graphical Behavior of Polynomials at Horizontal Intercepts

If a polynomial contains a factor of the form $(x-h)^p$, the behavior near the horizontal intercept h is determined by the power on the factor.

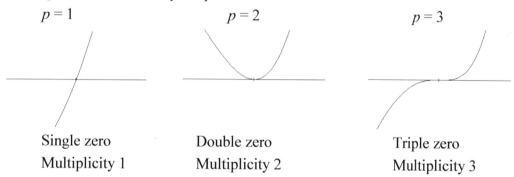

$p=1$	$p=2$	$p=3$
Single zero	Double zero	Triple zero
Multiplicity 1	Multiplicity 2	Multiplicity 3

For higher even powers 4,6,8 etc.... the graph will still bounce off the horizontal axis but the graph will appear flatter with each increasing even power as it approaches and leaves the axis.

For higher odd powers, 5,7,9 etc... the graph will still pass through the horizontal axis but the graph will appear flatter with each increasing odd power as it approaches and leaves the axis.

Example 4

Sketch a graph of $f(x) = -2(x+3)^2(x-5)$.

This graph has two horizontal intercepts. At $x = -3$, the factor is squared, indicating the graph will bounce at this horizontal intercept. At $x = 5$, the factor is not squared, indicating the graph will pass through the axis at this intercept.

Additionally, we can see the leading term, if this polynomial were multiplied out, would be $-2x^3$, so the long-run behavior is that of a vertically reflected cubic, with the outputs decreasing as the inputs get large positive, and the inputs increasing as the inputs get large negative.

To sketch this we consider the following:
As $x \to -\infty$ the function $f(x) \to \infty$ so we know the graph starts in the 2nd quadrant and is decreasing toward the horizontal axis.

At $(-3, 0)$ the graph bounces off the horizontal axis and so the function must start increasing.

At $(0, 90)$ the graph crosses the vertical axis at the vertical intercept.

Somewhere after this point, the graph must turn back down or start decreasing toward the horizontal axis since the graph passes through the next intercept at $(5,0)$.

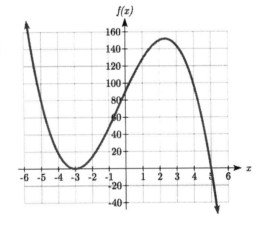

As $x \to \infty$ the function $f(x) \to -\infty$ so we know the graph continues to decrease and we can stop drawing the graph in the 4th quadrant.

Using technology we can verify the shape of the graph.

Try it Now

2. Given the function $g(x) = x^3 - x^2 - 6x$ use the methods that we have learned so far to find the vertical & horizontal intercepts, determine where the function is negative and positive, describe the long run behavior and sketch the graph without technology.

Solving Polynomial Inequalities

One application of our ability to find intercepts and sketch a graph of polynomials is the ability to solve polynomial inequalities. It is a very common question to ask when a function will be positive and negative. We can solve polynomial inequalities by either utilizing the graph, or by using test values.

Example 5

Solve $(x+3)(x+1)^2(x-4) > 0$

As with all inequalities, we start by solving the equality $(x+3)(x+1)^2(x-4) = 0$, which has solutions at $x = $ -3, -1, and 4. We know the function can only change from positive to negative at these values, so these divide the inputs into 4 intervals.

We could choose a test value in each interval and evaluate the function $f(x) = (x+3)(x+1)^2(x-4)$ at each test value to determine if the function is positive or negative in that interval

Interval	Test x in interval	$f($ test value$)$	>0 or <0?
$x < -3$	-4	72	> 0
$-3 < x < -1$	-2	-6	< 0
$-1 < x < 4$	0	-12	< 0
$x > 4$	5	288	> 0

On a number line this would look like:

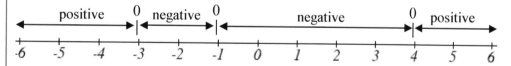

From our test values, we can determine this function is positive when $x < $ -3 or $x > 4$, or in interval notation, $(-\infty, -3) \cup (4, \infty)$

We could have also determined on which intervals the function was positive by sketching a graph of the function. We illustrate that technique in the next example

Example 6

Find the domain of the function $v(t) = \sqrt{6 - 5t - t^2}$.

A square root is only defined when the quantity we are taking the square root of, the quantity inside the square root, is zero or greater. Thus, the domain of this function will be when $6 - 5t - t^2 \geq 0$.

We start by solving the equality $6 - 5t - t^2 = 0$. While we could use the quadratic formula, this equation factors nicely to $(6 + t)(1 - t) = 0$, giving horizontal intercepts $t = 1$ and $t = -6$.

Sketching a graph of this quadratic will allow us to determine when it is positive.

From the graph we can see this function is positive for inputs between the intercepts. So $6 - 5t - t^2 \geq 0$ for $-6 \leq t \leq 1$, and this will be the domain of the $v(t)$ function.

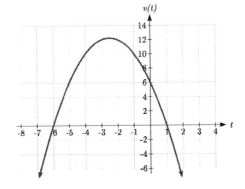

Writing Equations using Intercepts

Since a polynomial function written in factored form will have a horizontal intercept where each factor is equal to zero, we can form a function that will pass through a set of horizontal intercepts by introducing a corresponding set of factors.

Factored Form of Polynomials

If a polynomial has horizontal intercepts at $x = x_1, x_2, \ldots, x_n$, then the polynomial can be written in the factored form

$$f(x) = a(x - x_1)^{p_1}(x - x_2)^{p_2} \cdots (x - x_n)^{p_n}$$

where the powers p_i on each factor can be determined by the behavior of the graph at the corresponding intercept, and the stretch factor a can be determined given a value of the function other than the horizontal intercept.

Example 7

Write a formula for the polynomial function graphed here.

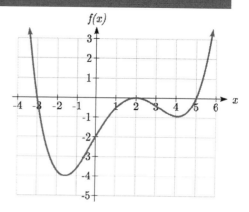

This graph has three horizontal intercepts: $x = -3$, 2, and 5. At $x = -3$ and 5 the graph passes through the axis, suggesting the corresponding factors of the polynomial will be linear. At $x = 2$ the graph bounces at the intercept, suggesting the corresponding factor of the polynomial will be 2nd degree (quadratic).

Together, this gives us:
$$f(x) = a(x+3)(x-2)^2(x-5)$$

To determine the stretch factor, we can utilize another point on the graph. Here, the vertical intercept appears to be (0,-2), so we can plug in those values to solve for a:
$$-2 = a(0+3)(0-2)^2(0-5)$$
$$-2 = -60a$$
$$a = \frac{1}{30}$$

The graphed polynomial appears to represent the function
$$f(x) = \frac{1}{30}(x+3)(x-2)^2(x-5).$$

Try it Now

3. Given the graph, write a formula for the function shown.

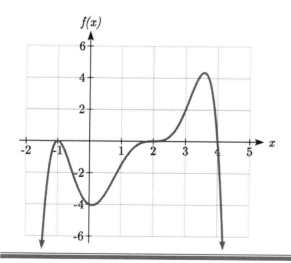

Estimating Extrema

With quadratics, we were able to algebraically find the maximum or minimum value of the function by finding the vertex. For general polynomials, finding these turning points is not possible without more advanced techniques from calculus. Even then, finding where extrema occur can still be algebraically challenging. For now, we will estimate the locations of turning points using technology to generate a graph.

Example 8

An open-top box is to be constructed by cutting out squares from each corner of a 14cm by 20cm sheet of plastic then folding up the sides. Find the size of squares that should be cut out to maximize the volume enclosed by the box.

We will start this problem by drawing a picture, labeling the width of the cut-out squares with a variable, w.

Notice that after a square is cut out from each end, it leaves a $(14-2w)$ cm by $(120-2w)$ cm rectangle for the base of the box, and the box will be w cm tall. This gives the volume:
$$V(w) = (14-2w)(20-2w)w = 280w - 68w^2 + 4w^3$$

Using technology to sketch a graph allows us to estimate the maximum value for the volume, restricted to reasonable values for w: values from 0 to 7.

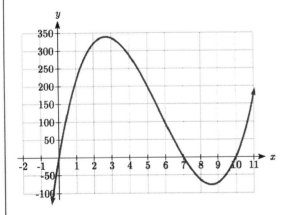

From this graph, we can estimate the maximum value is around 340, and occurs when the squares are about 2.75cm square. To improve this estimate, we could use advanced features of our technology, if available, or simply change our window to zoom in on our graph.

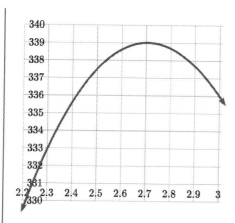

From this zoomed-in view, we can refine our estimate for the max volume to about 339, when the squares are 2.7cm square.

Try it Now

4. Use technology to find the maximum and minimum values on the interval [-1, 4] of the function $f(x) = -0.2(x-2)^3(x+1)^2(x-4)$.

Important Topics of this Section
Short Run Behavior
Intercepts (Horizontal & Vertical)
Methods to find Horizontal intercepts
Factoring Methods
Factored Forms
Technology
Graphical Behavior at intercepts
Single, Double and Triple zeros (or multiplicity 1, 2, and 3 behaviors)
Solving polynomial inequalities using test values & graphing techniques
Writing equations using intercepts
Estimating extrema

Try it Now Answers

1. Vertical intercept (0, 0). $0 = t^4 - 4t^2$ factors as $0 = t^2(t^2 - 4) = t^2(t - 2)(t + 2)$

   Horizontal intercepts (0, 0), (-2, 0), (2, 0)

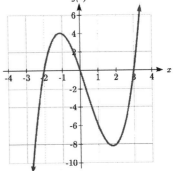

2. Vertical intercept (0, 0),

   Horizontal intercepts (-2, 0), (0, 0), (3, 0)

   The function is negative on $(-\infty, -2)$ and (0, 3)

   The function is positive on (-2, 0) and $(3, \infty)$

   The leading term is x^3 so as $x \to -\infty$, $g(x) \to -\infty$ and as

   $x \to \infty$, $g(x) \to \infty$

3. Double zero at $x=-1$, triple zero at $x=2$. Single zero at $x=4$.

   $f(x) = a(x-2)^3(x+1)^2(x-4)$. Substituting (0,-4) and solving for a,

   $f(x) = -\dfrac{1}{8}(x-2)^3(x+1)^2(x-4)$

4. The minimum occurs at approximately the point (0, -6.5), and the maximum occurs at
   approximately the point (3.5, 7).

Section 3.3 Exercises

Find the C and t intercepts of each function.

1. $C(t) = 2(t-4)(t+1)(t-6)$

2. $C(t) = 3(t+2)(t-3)(t+5)$

3. $C(t) = 4t(t-2)^2(t+1)$

4. $C(t) = 2t(t-3)(t+1)^2$

5. $C(t) = 2t^4 - 8t^3 + 6t^2$

6. $C(t) = 4t^4 + 12t^3 - 40t^2$

Use your calculator or other graphing technology to solve graphically for the zeros of the function.

7. $f(x) = x^3 - 7x^2 + 4x + 30$

8. $g(x) = x^3 - 6x^2 + x + 28$

Find the long run behavior of each function as $t \to \infty$ and $t \to -\infty$

9. $h(t) = 3(t-5)^3(t-3)^3(t-2)$

10. $k(t) = 2(t-3)^2(t+1)^3(t+2)$

11. $p(t) = -2t(t-1)(3-t)^2$

12. $q(t) = -4t(2-t)(t+1)^3$

Sketch a graph of each equation.

13. $f(x) = (x+3)^2(x-2)$

14. $g(x) = (x+4)(x-1)^2$

15. $h(x) = (x-1)^3(x+3)^2$

16. $k(x) = (x-3)^3(x-2)^2$

17. $m(x) = -2x(x-1)(x+3)$

18. $n(x) = -3x(x+2)(x-4)$

Solve each inequality.

19. $(x-3)(x-2)^2 > 0$

20. $(x-5)(x+1)^2 > 0$

21. $(x-1)(x+2)(x-3) < 0$

22. $(x-4)(x+3)(x+6) < 0$

Find the domain of each function.

23. $f(x) = \sqrt{-42 + 19x - 2x^2}$

24. $g(x) = \sqrt{28 - 17x - 3x^2}$

25. $h(x) = \sqrt{4 - 5x + x^2}$

26. $k(x) = \sqrt{2 + 7x + 3x^2}$

27. $n(x) = \sqrt{(x-3)(x+2)^2}$

28. $m(x) = \sqrt{(x-1)^2(x+3)}$

29. $p(t) = \dfrac{1}{t^2 + 2t - 8}$

30. $q(t) = \dfrac{4}{x^2 - 4x - 5}$

Write an equation for a polynomial the given features.

31. Degree 3. Zeros at $x = -2$, $x = 1$, and $x = 3$. Vertical intercept at $(0, -4)$

32. Degree 3. Zeros at $x = -5$, $x = -2$, and $x = 1$. Vertical intercept at $(0, 6)$

33. Degree 5. Roots of multiplicity 2 at $x = 3$ and $x = 1$, and a root of multiplicity 1 at $x = -3$. Vertical intercept at $(0, 9)$

34. Degree 4. Root of multiplicity 2 at $x = 4$, and a roots of multiplicity 1 at $x = 1$ and $x = -2$. Vertical intercept at $(0, -3)$

35. Degree 5. Double zero at $x = 1$, and triple zero at $x = 3$. Passes through the point $(2, 15)$

36. Degree 5. Single zero at $x = -2$ and $x = 3$, and triple zero at $x = 1$. Passes through the point $(2, 4)$

Write a formula for each polynomial function graphed.

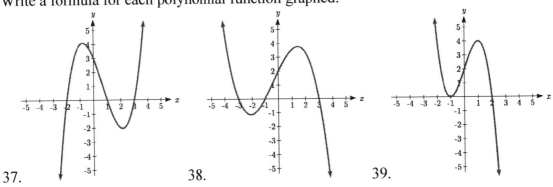

37.   38.   39.

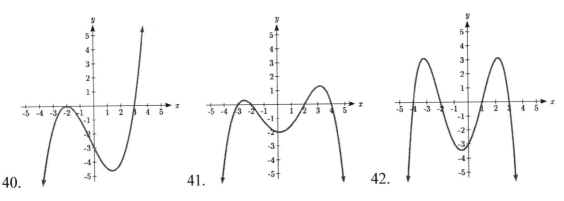

40.   41.   42.

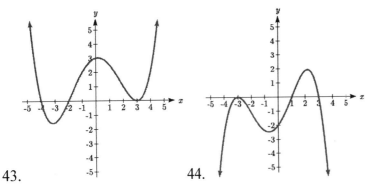

43.   44.

Write a formula for each polynomial function graphed.

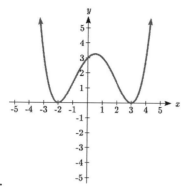

45.

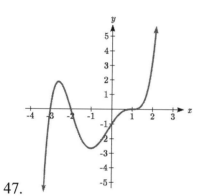

47.

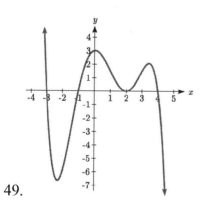

49.

46.

48.

50.

51. A rectangle is inscribed with its base on the x axis and its upper corners on the parabola $y = 5 - x^2$. What are the dimensions of such a rectangle that has the greatest possible area?

52. A rectangle is inscribed with its base on the x axis and its upper corners on the curve $y = 16 - x^4$. What are the dimensions of such a rectangle that has the greatest possible area?

Section 3.4 Factor Theorem and Remainder Theorem

In the last section, we limited ourselves to finding the intercepts, or zeros, of polynomials that factored simply, or we turned to technology. In this section, we will look at algebraic techniques for finding the zeros of polynomials like $h(t) = t^3 + 4t^2 + t - 6$.

Long Division

In the last section we saw that we could write a polynomial as a product of factors, each corresponding to a horizontal intercept. If we knew that $x = 2$ was an intercept of the polynomial $x^3 + 4x^2 - 5x - 14$, we might guess that the polynomial could be factored as $x^3 + 4x^2 - 5x - 14 = (x - 2)(\text{something})$. To find that "something," we can use polynomial division.

Example 1

Divide $x^3 + 4x^2 - 5x - 14$ by $x - 2$

Start by writing the problem out in long division form

$$x - 2 \overline{)x^3 + 4x^2 - 5x - 14}$$

Now we divide the leading terms: $x^3 \div x = x^2$. It is best to align it above the same-powered term in the dividend. Now, multiply that x^2 by $x - 2$ and write the result below the dividend.

$$\begin{array}{r} x^2 \\ x-2 \overline{)x^3 + 4x^2 - 5x - 14} \\ x^3 - 2x^2 \end{array}$$ Now subtract that expression from the dividend.

$$\begin{array}{r} x^2 \\ x-2 \overline{)x^3 + 4x^2 - 5x - 14} \\ -(x^3 - 2x^2) \\ \hline 6x^2 - 5x - 14 \end{array}$$

Again, divide the leading term of the remainder by the leading term of the divisor. $6x^2 \div x = 6x$. We add this to the result, multiply $6x$ by $x - 2$, and subtract.

$$\begin{array}{r}
x^2 + 6x \\
x - 2 \overline{\smash{)}\ x^3 + 4x^2 - 5x - 14} \\
\underline{-\left(x^3 - 2x^2\right)} \\
6x^2 - 5x - 14 \\
\underline{-\left(6x^2 - 12x\right)} \\
7x - 14
\end{array}$$

Repeat the process one last time.

$$\begin{array}{r}
x^2 + 6x + 7 \\
x - 2 \overline{\smash{)}\ x^3 + 4x^2 - 5x - 14} \\
\underline{-\left(x^3 - 2x^2\right)} \\
6x^2 - 5x - 14 \\
\underline{-\left(6x^2 - 12x\right)} \\
7x - 14 \\
\underline{-\left(7x - 14\right)} \\
0
\end{array}$$

This tells us $x^3 + 4x^2 - 5x - 14$ divided by $x - 2$ is $x^2 + 6x + 7$, with a remainder of zero. This also means that we can factor $x^3 + 4x^2 - 5x - 14$ as $(x - 2)(x^2 + 6x + 7)$.

This gives us a way to find the intercepts of this polynomial.

Example 2

Find the horizontal intercepts of $h(x) = x^3 + 4x^2 - 5x - 14$.

To find the horizontal intercepts, we need to solve $h(x) = 0$. From the previous example, we know the function can be factored as $h(x) = (x - 2)(x^2 + 6x + 7)$.

$h(x) = (x - 2)(x^2 + 6x + 7) = 0$ when $x = 2$ or when $x^2 + 6x + 7 = 0$. This doesn't factor nicely, but we could use the quadratic formula to find the remaining two zeros.

$$x = \frac{-6 \pm \sqrt{6^2 - 4(1)(7)}}{2(1)} = -3 \pm \sqrt{2} \, .$$

The horizontal intercepts will be at $(2,0)$, $\left(-3 - \sqrt{2}, 0\right)$, and $\left(-3 + \sqrt{2}, 0\right)$.

Try it Now
1. Divide $2x^3 - 7x + 3$ by $x + 3$ using long division.

The Factor and Remainder Theorems

When we divide a polynomial, $p(x)$ by some divisor polynomial $d(x)$, we will get a quotient polynomial $q(x)$ and possibly a remainder $r(x)$. In other words,
$p(x) = d(x)q(x) + r(x)$.

Because of the division, the remainder will either be zero, or a polynomial of lower degree than $d(x)$. Because of this, if we divide a polynomial by a term of the form $x - c$, then the remainder will be zero or a constant.

If $p(x) = (x - c)q(x) + r$, then $p(c) = (c - c)q(c) + r = 0 + r = r$, which establishes the Remainder Theorem.

> ### The Remainder Theorem
> If $p(x)$ is a polynomial of degree 1 or greater and c is a real number, then when $p(x)$ is divided by $x - c$, the remainder is $p(c)$.

If $x - c$ is a factor of the polynomial p, then $p(x) = (x - c)q(x)$ for some polynomial q. Then $p(c) = (c - c)q(c) = 0$, showing c is a zero of the polynomial. This shouldn't surprise us - we already knew that if the polynomial factors it reveals the roots.

If $p(c) = 0$, then the remainder theorem tells us that if p is divided by $x - c$, then the remainder will be zero, which means $x - c$ is a factor of p.

> ### The Factor Theorem
> If $p(x)$ is a nonzero polynomial, then the real number c is a zero of $p(x)$ if and only if $x - c$ is a factor of $p(x)$.

Synthetic Division

Since dividing by $x - c$ is a way to check if a number is a zero of the polynomial, it would be nice to have a faster way to divide by $x - c$ than having to use long division every time. Happily, quicker ways have been discovered.

Let's look back at the long division we did in Example 1 and try to streamline it. First, let's change all the subtractions into additions by distributing through the negatives.

$$
\begin{array}{r}
x^2+6x+7 \\
x-2\overline{\smash{\big)}\ x^3+4x^2-5x-14} \\
\underline{-x^3+2x^2} \\
6x^2-5x-14 \\
\underline{-6x^2+12x} \\
7x-14 \\
\underline{-7x+14} \\
0
\end{array}
$$

Next, observe that the terms $-x^3$, $-6x^2$, and $-7x$ are the exact opposite of the terms above them. The algorithm we use ensures this is always the case, so we can omit them without losing any information. Also note that the terms we 'bring down' (namely the $-5x$ and -14) aren't really necessary to recopy, so we omit them, too.

$$
\begin{array}{r}
x^2+6x+7 \\
x-2\overline{\smash{\big)}\ x^3+4x^2-5x-14} \\
\underline{2x^2} \\
6x^2 \\
\underline{12x} \\
7x \\
\underline{14} \\
0
\end{array}
$$

Now, let's move things up a bit and, for reasons which will become clear in a moment, copy the x^3 into the last row.

$$
\begin{array}{r}
x^2+6x+7 \\
x-2\overline{\smash{\big)}x^3+4x^2-5x-14} \\
\underline{2x^2\ \ 12x\ \ \ 14} \\
x^3\ \ \ 6x^2\ \ 7x\ \ \ 0
\end{array}
$$

Note that by arranging things in this manner, each term in the last row is obtained by adding the two terms above it. Notice also that the quotient polynomial can be obtained by dividing each of the first three terms in the last row by x and adding the results. If you take the time to work back through the original division problem, you will find that this is exactly the way we determined the quotient polynomial.

This means that we no longer need to write the quotient polynomial down, nor the x in the divisor, to determine our answer.

$$
\begin{array}{r}
x-2 \overline{\smash{\big)}\, x^3 + 4x^2 - 5x - 14} \\
\underline{2x^2 \quad 12x \quad 14} \\
x^3 \quad 6x^2 \quad 7x \quad 0
\end{array}
$$

We've streamlined things quite a bit so far, but we can still do more. Let's take a moment to remind ourselves where the $2x^2$, $12x$ and 14 came from in the second row. Each of these terms was obtained by multiplying the terms in the quotient, x^2, $6x$ and 7, respectively, by the -2 in $x - 2$, then by -1 when we changed the subtraction to addition. Multiplying by -2 then by -1 is the same as multiplying by 2, so we replace the -2 in the divisor by 2. Furthermore, the coefficients of the quotient polynomial match the coefficients of the first three terms in the last row, so we now take the plunge and write only the coefficients of the terms to get

$$
\begin{array}{r}
2\,|\quad 1 \quad\ 4 \quad -5 \quad -14 \\
\underline{\quad\quad\ 2 \quad\ 12 \quad\ 14} \\
1 \quad\ 6 \quad\ 7 \quad\ \ 0
\end{array}
$$

We have constructed a **synthetic division** tableau for this polynomial division problem. Let's re-work our division problem using this tableau to see how it greatly streamlines the division process. To divide $x^3 + 4x^2 - 5x - 14$ by $x - 2$, we write 2 in the place of the divisor and the coefficients of $x^3 + 4x^2 - 5x - 14$ in for the dividend. Then "bring down" the first coefficient of the dividend.

$$
\begin{array}{r}
2\,|\quad 1 \quad\ 4 \quad -5 \quad -14 \\
\underline{} \\

\end{array}
\qquad
\begin{array}{r}
2\,|\quad 1 \quad\ 4 \quad -5 \quad -14 \\
\downarrow \\
\underline{} \\
1
\end{array}
$$

Next, take the 2 from the divisor and multiply by the 1 that was "brought down" to get 2. Write this underneath the 4, then add to get 6.

$$
\begin{array}{r}
2\,|\quad 1 \quad\ 4 \quad -5 \quad -14 \\
\downarrow \quad 2 \\
\underline{} \\
1
\end{array}
\qquad
\begin{array}{r}
2\,|\quad 1 \quad\ 4 \quad -5 \quad -14 \\
\downarrow \quad 2 \\
\underline{} \\
1 \quad 6
\end{array}
$$

Now take the 2 from the divisor times the 6 to get 12, and add it to the -5 to get 7.

$$
\begin{array}{r}
2\,|\quad 1 \quad\ 4 \quad -5 \quad -14 \\
\downarrow \quad 2 \quad 12 \\
\underline{} \\
1 \quad 6
\end{array}
\qquad
\begin{array}{r}
2\,|\quad 1 \quad\ 4 \quad -5 \quad -14 \\
\downarrow \quad 2 \quad 12 \\
\underline{} \\
1 \quad 6 \quad 7
\end{array}
$$

Finally, take the 2 in the divisor times the 7 to get 14, and add it to the -14 to get 0.

$$
\begin{array}{r|rrrr}
2 & 1 & 4 & -5 & -14 \\
  & \downarrow & 2 & 12 & 14 \\
\hline
  & 1 & 6 & 7
\end{array}
\qquad
\begin{array}{r|rrrr}
2 & 1 & 4 & -5 & -14 \\
  & \downarrow & 2 & 12 & 14 \\
\hline
  & 1 & 6 & 7 & \boxed{0}
\end{array}
$$

The first three numbers in the last row of our tableau are the coefficients of the quotient polynomial. Remember, we started with a third degree polynomial and divided by a first degree polynomial, so the quotient is a second degree polynomial. Hence the quotient is $x^2 + 6x + 7$. The number in the box is the remainder. Synthetic division is our tool of choice for dividing polynomials by divisors of the form $x - c$. It is important to note that it works only for these kinds of divisors. Also take note that when a polynomial (of degree at least 1) is divided by $x - c$, the result will be a polynomial of exactly one less degree. Finally, it is worth the time to trace each step in synthetic division back to its corresponding step in long division.

Example 3

Use synthetic division to divide $5x^3 - 2x^2 + 1$ by $x - 3$.

When setting up the synthetic division tableau, we need to enter 0 for the coefficient of x in the dividend. Doing so gives

$$
\begin{array}{r|rrrr}
3 & 5 & -2 & 0 & 1 \\
  & \downarrow & 15 & 39 & 117 \\
\hline
  & 5 & 13 & 39 & \boxed{118}
\end{array}
$$

Since the dividend was a third degree polynomial, the quotient is a quadratic polynomial with coefficients 5, 13 and 39. Our quotient is $q(x) = 5x^2 + 13x + 39$ and the remainder is $r(x) = 118$. This means
$5x^3 - 2x^2 + 1 = (x - 3)(5x^2 + 13x + 39) + 118$.

It also means that $x - 3$ is *not* a factor of $5x^3 - 2x^2 + 1$.

Example 4

Divide $x^3 + 8$ by $x + 2$

For this division, we rewrite $x + 2$ as $x - (-2)$ and proceed as before.

$$
\begin{array}{r|rrrr}
-2 & 1 & 0 & 0 & 8 \\
   & \downarrow & -2 & 4 & -8 \\
\hline
   & 1 & -2 & 4 & \boxed{0}
\end{array}
$$

The quotient is $x^2 - 2x + 4$ and the remainder is zero. Since the remainder is zero, $x + 2$ is a factor of $x^3 + 8$.

$$x^3 + 8 = (x + 2)\left(x^2 - 2x + 4\right)$$

Try it Now

2. Divide $4x^4 - 8x^2 - 5x$ by $x - 3$ using synthetic division.

Using this process allows us to find the real zeros of polynomials, presuming we can figure out at least one root. We'll explore how to do that in the next section.

Example 5

The polynomial $p(x) = 4x^4 - 4x^3 - 11x^2 + 12x - 3$ has a horizontal intercept at $x = \dfrac{1}{2}$ with multiplicity 2. Find the other intercepts of $p(x)$.

Since $x = \dfrac{1}{2}$ is an intercept with multiplicity 2, then $x - \dfrac{1}{2}$ is a factor twice. Use synthetic division to divide by $x - \dfrac{1}{2}$ twice.

```
1/2 |  4   -4   -11   12   -3
     ↓       2    -1   -6    3
       4    -2    -1   -6 | 0

1/2 |  4   -2   -1   -6
     ↓       2    0   -6
       4    0  -12 | 0
```

From the first division, we get $4x^4 - 4x^3 - 11x^2 + 12x - 3 = \left(x - \dfrac{1}{2}\right)\left(4x^3 - 2x^2 - x - 6\right)$

The second division tells us

$$4x^4 - 4x^3 - 11x^2 + 12x - 3 = \left(x - \dfrac{1}{2}\right)\left(x - \dfrac{1}{2}\right)\left(4x^2 - 12\right).$$

To find the remaining intercepts, we set $4x^2 - 12 = 0$ and get $x = \pm\sqrt{3}$.

Note this also means $4x^4 - 4x^3 - 11x^2 + 12x - 3 = 4\left(x - \dfrac{1}{2}\right)\left(x - \dfrac{1}{2}\right)\left(x - \sqrt{3}\right)\left(x + \sqrt{3}\right)$.

Important Topics of this Section
Long division of polynomials
Remainder Theorem
Factor Theorem
Synthetic division of polynomials

Try it Now Answers

1.

$$2x^2 - 6x + 11$$

$$x + 3 \overline{\smash{)}\ 2x^3 + 0x^2 - 7x + 3}$$

$$\underline{-\left(2x^3 + 6x^2\right)}$$

$$-6x^2 - 7x + 3$$

$$\underline{-\left(-6x^2 - 18x\right)}$$

$$11x + 3$$

$$\underline{-\left(11x + 33\right)}$$

$$-30$$

The quotient is $2x^2 - 6x + 11$ with remainder -30.

2.

$$
\begin{array}{r|rrrrr}
3 & 4 & 0 & -8 & -5 & 0 \\
  &   & 12 & 36 & 84 & 237 \\
\hline
  & 4 & 12 & 28 & 79 & \boxed{237}
\end{array}
$$

$4x^4 - 8x^2 - 5x$ divided by $x - 3$ is $4x^3 + 12x^2 + 28x + 79$ with remainder 237

Section 3.4 Exercises

Use polynomial long division to perform the indicated division.

1. $\left(4x^2 + 3x - 1\right) \div (x - 3)$

2. $\left(2x^3 - x + 1\right) \div \left(x^2 + x + 1\right)$

3. $\left(5x^4 - 3x^3 + 2x^2 - 1\right) \div \left(x^2 + 4\right)$

4. $\left(-x^5 + 7x^3 - x\right) \div \left(x^3 - x^2 + 1\right)$

5. $\left(9x^3 + 5\right) \div (2x - 3)$

6. $\left(4x^2 - x - 23\right) \div \left(x^2 - 1\right)$

Use synthetic division to perform the indicated division.

7. $\left(3x^2 - 2x + 1\right) \div (x - 1)$

8. $\left(x^2 - 5\right) \div (x - 5)$

9. $\left(3 - 4x - 2x^2\right) \div (x + 1)$

10. $\left(4x^2 - 5x + 3\right) \div (x + 3)$

11. $\left(x^3 + 8\right) \div (x + 2)$

12. $\left(4x^3 + 2x - 3\right) \div (x - 3)$

13. $\left(18x^2 - 15x - 25\right) \div \left(x - \dfrac{5}{3}\right)$

14. $\left(4x^2 - 1\right) \div \left(x - \dfrac{1}{2}\right)$

15. $\left(2x^3 + x^2 + 2x + 1\right) \div \left(x + \dfrac{1}{2}\right)$

16. $\left(3x^3 - x + 4\right) \div \left(x - \dfrac{2}{3}\right)$

17. $\left(2x^3 - 3x + 1\right) \div \left(x - \dfrac{1}{2}\right)$

18. $\left(4x^4 - 12x^3 + 13x^2 - 12x + 9\right) \div \left(x - \dfrac{3}{2}\right)$

19. $\left(x^4 - 6x^2 + 9\right) \div \left(x - \sqrt{3}\right)$

20. $\left(x^6 - 6x^4 + 12x^2 - 8\right) \div \left(x + \sqrt{2}\right)$

Below you are given a polynomial and one of its zeros. Use the techniques in this section to find the rest of the real zeros and factor the polynomial.

21. $x^3 - 6x^2 + 11x - 6$, $c = 1$

22. $x^3 - 24x^2 + 192x - 512$, $c = 8$

23. $3x^3 + 4x^2 - x - 2$, $c = \dfrac{2}{3}$

24. $2x^3 - 3x^2 - 11x + 6$, $c = \dfrac{1}{2}$

25. $x^3 + 2x^2 - 3x - 6$, $c = -2$

26. $2x^3 - x^2 - 10x + 5$, $c = \dfrac{1}{2}$

27. $4x^4 - 28x^3 + 61x^2 - 42x + 9$, $c = \dfrac{1}{2}$ is a zero of multiplicity 2

28. $x^5 + 2x^4 - 12x^3 - 38x^2 - 37x - 12$, $c = -1$ is a zero of multiplicity 3

Section 3.5 Real Zeros of Polynomials

In the last section, we saw how to determine if a real number was a zero of a polynomial. In this section, we will learn how to find good candidates to test using synthetic division. In the days before graphing technology was commonplace, mathematicians discovered a lot of clever tricks for determining the likely locations of zeros. Technology has provided a much simpler approach to narrow down potential candidates, but it is not always sufficient by itself. For example, the function shown to the right does not have any clear intercepts.

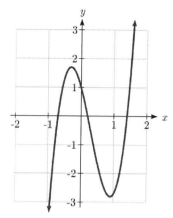

There are two results that can help us identify where the zeros of a polynomial are. The first gives us an interval on which all the real zeros of a polynomial can be found.

Cauchy's Bound

Given a polynomial $f(x) = a_n x^n + a_{n-1} x^{n-1} + \cdots + a_1 x + a_0$, let M be the largest of the coefficients in absolute value. Then all the real zeros of $f(x)$ lie in the interval

$$\left[-\frac{M}{|a_n|} - 1, \quad \frac{M}{|a_n|} + 1 \right]$$

Example 1

Let $f(x) = 2x^4 + 4x^3 - x^2 - 6x - 3$. Determine an interval which contains all the real zeros of f.

To find the M from Cauchy's Bound, we take the absolute value of the coefficients and pick the largest, in this case $|-6| = 6$. Divide this by the absolute value of the leading coefficient, 2, to get 3. All the real zeros of f lie in the interval

$$\left[-\frac{6}{|2|} - 1, \quad \frac{6}{|2|} + 1 \right] = \left[-3 - 1, \quad 3 + 1 \right] = [-4, 4].$$

Knowing this bound can be very helpful when using a graphing calculator, since we can use it to set the display bounds. This helps avoid missing a zero because it is graphed outside of the viewing window.

1. Determine an interval which contains all the real zeros of $f(x) = 3x^3 - 12x^2 + 6x - 8$

Now that we know *where* we can find the real zeros, we still need a list of *possible* real zeros. The Rational Roots Theorem provides us a list of potential integer and rational zeros.

Rational Roots Theorem

Given a polynomial $f(x) = a_n x^n + a_{n-1} x^{n-1} + \cdots + a_1 x + a_0$ with integer coefficients, if r is a rational zero of f, then r is of the form $r = \pm \dfrac{p}{q}$, where p is a factor of the constant term a_0, and q is a factor of the leading coefficient, a_n.

This gives us a list of numbers to try in our synthetic division, which is a nicer place to start than simply guessing. If none of the numbers in the list are zeros, then either the polynomial has no real zeros at all, or all the real zeros are irrational numbers.

Example 2

Let $f(x) = 2x^4 + 4x^3 - x^2 - 6x - 3$. Use the Rational Roots Theorem to list all the possible rational zeros of *f(x)*.

To generate a complete list of rational zeros, we need to take each of the factors of the constant term, $a_0 = -3$, and divide them by each of the factors of the leading coefficient $a_4 = 2$. The factors of -3 are ± 1 and ± 3. Since the Rational Roots Theorem tacks on a $\pm$ anyway, for the moment, we consider only the positive factors 1 and 3. The factors of 2 are 1 and 2, so the Rational Roots Theorem gives the list

$$\left\{ \pm \frac{1}{1}, \pm \frac{1}{2}, \pm \frac{3}{1}, \pm \frac{3}{2} \right\}, \text{ or } \left\{ \pm 1, \pm \frac{1}{2}, \pm 3, \pm \frac{3}{2} \right\}$$

Now we can use synthetic division to test these possible zeros. To narrow the list first, we could use graphing technology to help us identify some good possibilities.

Example 3

Find the horizontal intercepts of $f(x) = 2x^4 + 4x^3 - x^2 - 6x - 3$.

From Example 1, we know that the real zeros lie in the interval [-4, 4]. Using a graphing calculator, we could set the window accordingly and get the graph below.

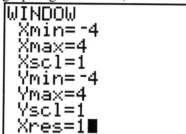
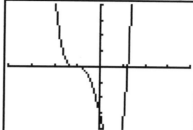

In Example 2, we learned that any rational zero must be on the list $\left\{\pm 1, \pm \dfrac{1}{2}, \pm 3, \pm \dfrac{3}{2}\right\}$.

From the graph, it looks like -1 is a good possibility, so we try that using synthetic division.

$$
\begin{array}{r|rrrrr}
-1 & 2 & 4 & -1 & -6 & -3 \\
 & \downarrow & -2 & -2 & 3 & 3 \\
\hline
 & 2 & 2 & -3 & -3 & \boxed{0}
\end{array}
$$

Success! Remembering that f was a fourth degree polynomial, we know that our quotient is a third degree polynomial. If we can do one more successful division, we will have knocked the quotient down to a quadratic, and, if all else fails, we can use the quadratic formula to find the last two zeros. Since there seems to be no other rational zeros to try, we continue with -1. Also, the shape of the crossing at $x = -1$ leads us to wonder if the zero $x = -1$ has multiplicity 3.

$$
\begin{array}{r|rrrr}
-1 & 2 & 2 & -3 & -3 \\
 & \downarrow & -2 & 0 & 3 \\
\hline
 & 2 & 0 & -3 & \boxed{0}
\end{array}
$$

Success again! Our quotient polynomial is now $2x^2 - 3$. Setting this to zero gives $2x^2 - 3 = 0$, giving $x = \pm\sqrt{\dfrac{3}{2}} = \pm\dfrac{\sqrt{6}}{2}$. Since a fourth degree polynomial can have at most four zeros, including multiplicities, then the intercept $x = -1$ must only have multiplicity 2, which we had found through division, and not 3 as we had guessed.

It is interesting to note that we could greatly improve on the graph of $y = f(x)$ in the previous example given to us by the calculator. For instance, from our determination of the zeros of f and their multiplicities, we know the graph crosses at $x = -\dfrac{\sqrt{6}}{2} \approx -1.22$ then turns back upwards to touch the x–axis at $x = -1$. This tells us that, despite what the calculator showed us the first time, there is a relative maximum occurring at $x = -1$ and not a "flattened crossing" as we originally believed.

After resizing the window, we see not only the relative maximum but also a relative minimum just to the left of $x = -1$.

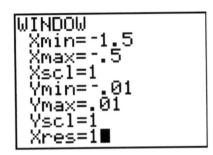
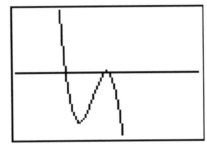

In this case, mathematics helped reveal something that was hidden in the initial graph.

Example 4

Find the real zeros of $f(x) = 4x^3 - 10x^2 - 2x + 2$.

Cauchy's Bound tells us that the real zeros lie in the interval $\left[-\dfrac{10}{|4|} - 1, \ \dfrac{10}{|4|} + 1 \right] = [-3.5, 3.5]$.

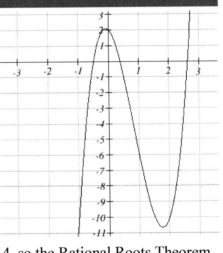

Graphing on this interval reveals no clear integer zeros. Turning to the rational roots theorem, we need to take each of the factors of the constant term, $a_0 = 2$, and divide them by each of the factors of the leading coefficient $a_3 = 4$. The factors of 2 are 1 and 2. The factors of 4 are 1, 2, and 4, so the Rational Roots Theorem gives the list

$$\left\{ \pm\frac{1}{1}, \pm\frac{1}{2}, \pm\frac{1}{4}, \pm\frac{2}{1}, \pm\frac{2}{2}, \pm\frac{2}{4} \right\}, \text{ or } \left\{ \pm 1, \pm\frac{1}{2}, \pm\frac{1}{4}, \pm 2 \right\}$$

The two most likely candidates are $\pm\dfrac{1}{2}$.

Trying $\dfrac{1}{2}$,

$$
\begin{array}{r|rrrr}
1/2 & 4 & -10 & -2 & 2 \\
& \downarrow & 2 & -4 & -3 \\
\hline
& 4 & -8 & -6 & \boxed{-1}
\end{array}
$$

The remainder is not zero, so this is not a zero. Trying $-\dfrac{1}{2}$,

$$
\begin{array}{r|rrrr}
-1/2 & 4 & -10 & -2 & 2 \\
& \downarrow & -2 & 6 & -2 \\
\hline
& 4 & -12 & 4 & \boxed{0}
\end{array}
$$

Success! This tells us $4x^3 - 10x^2 - 2x + 2 = \left(x + \dfrac{1}{2}\right)\left(4x^2 - 12x + 4\right)$, and that the graph

has a horizontal intercept at $x = -\dfrac{1}{2}$.

To find the remaining two intercepts, we can use the quadratic equation, setting
$4x^2 - 12x + 4 = 0$. First, we might pull out the common factor, $4\left(x^2 - 3x + 1\right) = 0$.

$$
x = \frac{3 \pm \sqrt{(-3)^2 - 4(1)(1)}}{2(1)} = \frac{3 \pm \sqrt{5}}{2} \approx 2.618, \ 0.382
$$

Try it Now

2. Find the real zeros of $f(x) = 3x^3 - x^2 - 6x + 2$

Important Topics of this Section
Cauchy's Bound for all real zeros of a polynomial
Rational Roots Theorem
Finding real zeros of a polynomial

Try it Now Answers

1. The maximum coefficient in absolute value is 12. Cauchy's Bound for all real zeros is

$$\left[-\frac{12}{|3|}-1, \quad \frac{12}{|3|}+1 \right]=[-5,5]$$

2. Cauchy's Bound tells us the zeros lie in the interval $\left[-\frac{6}{|3|}-1, \quad \frac{6}{|3|}+1 \right]=[-3,3]$.

The rational roots theorem tells us the possible rational zeros of the polynomial are on the list

$$\left\{ \pm\frac{1}{1},\pm\frac{1}{3},\pm\frac{2}{1},\pm\frac{2}{3} \right\}=\left\{ \pm 1,\pm\frac{1}{3},\pm 2,\pm\frac{2}{3} \right\}.$$

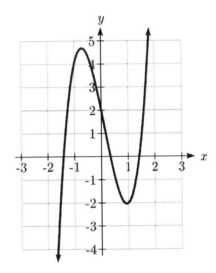

Looking at a graph, the only likely candidate is $\frac{1}{3}$

Using synthetic division,

$$
\begin{array}{r|rrrr}
1/3 & 3 & -1 & -6 & 2 \\
 & \downarrow & 1 & 0 & -2 \\
\hline
 & 3 & 0 & -6 & \boxed{0}
\end{array}
$$

$$3x^3-x^2-6x+2=\left(x-\frac{1}{3} \right)\!\left(3x^2-6\right)=3\left(x-\frac{1}{3} \right)\!\left(x^2-2\right).$$

Solving $x^2-2=0$ gives zeros $x=\pm\sqrt{2}$.

The real zeros of the polynomial are $x=\sqrt{2},-\sqrt{2},\frac{1}{3}$.

Section 3.5 Exercises

For each of the following polynomials, use Cauchy's Bound to find an interval containing all the real zeros, then use Rational Roots Theorem to make a list of possible rational zeros.

1. $f(x) = x^3 - 2x^2 - 5x + 6$

2. $f(x) = x^4 + 2x^3 - 12x^2 - 40x - 32$

3. $f(x) = x^4 - 9x^2 - 4x + 12$

4. $f(x) = x^3 + 4x^2 - 11x + 6$

5. $f(x) = x^3 - 7x^2 + x - 7$

6. $f(x) = -2x^3 + 19x^2 - 49x + 20$

7. $f(x) = -17x^3 + 5x^2 + 34x - 10$

8. $f(x) = 36x^4 - 12x^3 - 11x^2 + 2x + 1$

9. $f(x) = 3x^3 + 3x^2 - 11x - 10$

10. $f(x) = 2x^4 + x^3 - 7x^2 - 3x + 3$

Find the real zeros of each polynomial.

11. $f(x) = x^3 - 2x^2 - 5x + 6$

12. $f(x) = x^4 + 2x^3 - 12x^2 - 40x - 32$

13. $f(x) = x^4 - 9x^2 - 4x + 12$

14. $f(x) = x^3 + 4x^2 - 11x + 6$

15. $f(x) = x^3 - 7x^2 + x - 7$

16. $f(x) = -2x^3 + 19x^2 - 49x + 20$

17. $f(x) = -17x^3 + 5x^2 + 34x - 10$

18. $f(x) = 36x^4 - 12x^3 - 11x^2 + 2x + 1$

19. $f(x) = 3x^3 + 3x^2 - 11x - 10$

20. $f(x) = 2x^4 + x^3 - 7x^2 - 3x + 3$

21. $f(x) = 9x^3 - 5x^2 - x$

22. $f(x) = 6x^4 - 5x^3 - 9x^2$

23. $f(x) = x^4 + 2x^2 - 15$

24. $f(x) = x^4 - 9x^2 + 14$

25. $f(x) = 3x^4 - 14x^2 - 5$

26. $f(x) = 2x^4 - 7x^2 + 6$

27. $f(x) = x^6 - 3x^3 - 10$

28. $f(x) = 2x^6 - 9x^3 + 10$

29. $f(x) = x^5 - 2x^4 - 4x + 8$

30. $f(x) = 2x^5 + 3x^4 - 18x - 27$

31. $f(x) = x^5 - 60x^3 - 80x^2 + 960x + 2304$

32. $f(x) = 25x^5 - 105x^4 + 174x^3 - 142x^2 + 57x - 9$

Section 3.6 Complex Zeros

When finding the zeros of polynomials, at some point you're faced with the problem $x^2 = -1$. While there are clearly no real numbers that are solutions to this equation, leaving things there has a certain feel of incompleteness. To address that, we will need utilize the imaginary unit, i.

Imaginary Number i

The most basic complex number is i, defined to be $i = \sqrt{-1}$, commonly called an **imaginary number**. Any real multiple of i is also an imaginary number.

Example 1

Simplify $\sqrt{-9}$.

We can separate $\sqrt{-9}$ as $\sqrt{9}\sqrt{-1}$. We can take the square root of 9, and write the square root of -1 as i.

$$\sqrt{-9} = \sqrt{9}\sqrt{-1} = 3i$$

A complex number is the sum of a real number and an imaginary number.

Complex Number

A **complex number** is a number $z = a + bi$, where a and b are real numbers

a is the real part of the complex number

b is the imaginary part of the complex number

$i = \sqrt{-1}$

Arithmetic on Complex Numbers

Before we dive into the more complicated uses of complex numbers, let's make sure we remember the basic arithmetic involved. To add or subtract complex numbers, we simply add the like terms, combining the real parts and combining the imaginary parts.

Example 3

Add $3 - 4i$ and $2 + 5i$.

Adding $(3 - 4i) + (2 + 5i)$, we add the real parts and the imaginary parts
$3 + 2 - 4i + 5i$
$5 + i$

Try it Now
1. Subtract $2 + 5i$ from $3 - 4i$.

We can also multiply and divide complex numbers.

Example 4

Multiply: $4(2 + 5i)$.

To multiply the complex number by a real number, we simply distribute as we would when multiplying polynomials.

$4(2 + 5i)$
$= 4 \cdot 2 + 4 \cdot 5i$
$= 8 + 20i$

Example 5

Divide $\dfrac{(2 + 5i)}{(4 - i)}$.

To divide two complex numbers, we have to devise a way to write this as a complex number with a real part and an imaginary part.

We start this process by eliminating the complex number in the denominator. To do this, we multiply the numerator and denominator by a special complex number so that the result in the denominator is a real number. The number we need to multiply by is called the **complex conjugate**, in which the sign of the imaginary part is changed.

Here, $4 + i$ is the complex conjugate of $4 - i$. Of course, obeying our algebraic rules, we must multiply by $4 + i$ on both the top and bottom.
$\dfrac{(2 + 5i)}{(4 - i)} \cdot \dfrac{(4 + i)}{(4 + i)}$

To multiply two complex numbers, we expand the product as we would with polynomials (the process commonly called FOIL – "first outer inner last"). In the numerator:

$(2+5i)(4+i)$ Expand

$=8+20i+2i+5i^2$ Since $i=\sqrt{-1}$, $i^2=-1$

$=8+20i+2i+5(-1)$ Simplify

$=3+22i$

Following the same process to multiply the denominator

$(4-i)(4+i)$ Expand

$=(16-4i+4i-i^2)$ Since $i=\sqrt{-1}$, $i^2=-1$

$=(16-(-1))$

$=17$

Combining this we get $\dfrac{3+22i}{17}=\dfrac{3}{17}+\dfrac{22i}{17}$

Try it Now

2. Multiply $3-4i$ and $2+3i$.

In the last example, we used the conjugate of a complex number

Complex Conjugate
The **conjugate** of a complex number $a+bi$ is the number $a-bi$.
The notation commonly used for conjugation is a bar: $\overline{a+bi}=a-bi$

Complex Zeros of Polynomials

Complex numbers allow us a way to write solutions to quadratic equations that do not have real solutions.

Example 6

Find the zeros of $f(x) = x^2 - 2x + 5$.

Using the quadratic formula,

$$x = \frac{2 \pm \sqrt{(-2)^2 - 4(1)(5)}}{2(1)} = \frac{2 \pm \sqrt{-16}}{2} = \frac{2 \pm 4i}{2} = 1 \pm 2i.$$

Try it Now

3. Find the zeros of $f(x) = 2x^2 + 3x + 4$.

Two things are important to note. First, the zeros $1 + 2i$ and $1 - 2i$ are complex conjugates. This will always be the case when we find non-real zeros to a quadratic function with real coefficients.

Second, we could write $f(x) = x^2 - 2x + 5 = (x - (1 + 2i))(x - (1 - 2i))$ if we really wanted to, so the Factor and Remainder Theorems hold.

How do we know if a general polynomial has any complex zeros? We have seen examples of polynomials with no real zeros; can there be polynomials with no zeros at all? The answer to that last question, which comes from the Fundamental Theorem of Algebra, is "No."

Fundamental Theorem of Algebra

A non-constant polynomial f with real or complex coefficients will have at least one real or complex zero.

This theorem is an example of an "existence" theorem in mathematics. It guarantees the existence of at least one zero, but provides no algorithm to use for finding it.

Now suppose we have a polynomial $f(x)$ of degree n. The Fundamental Theorem of Algebra guarantees at least one zero z_1, then the Factor Theorem guarantees that f can be factored as $f(x) = (x - z_1)q_1(x)$, where the quotient $q_1(x)$ will be of degree $n-1$.

If this function is non-constant, than the Fundamental Theorem of Algebra applies to it, and we can find another zero. This can be repeated n times.

Complex Factorization Theorem

If f is a polynomial f with real or complex coefficients with degree $n \geq 1$, then f has exactly n real or complex zeros, counting multiplicities.

If $z_1, z_2, \ldots, z_k$ are the distinct zero of f with multiplicities $m_1, m_2, \ldots, m_k$ respectively, then $f(x) = a(x - z_1)^{m_1}(x - z_2)^{m_2} \cdots (x - z_k)^{m_k}$

Example 7

Find all the real and complex zeros of $f(x) = 12x^5 - 20x^4 + 19x^3 - 6x^2 - 2x + 1$.

Using the Rational Roots Theorem, the possible real rational roots are

$$\left\{ \pm\frac{1}{1}, \pm\frac{1}{2}, \pm\frac{1}{3}, \pm\frac{1}{4}, \pm\frac{1}{6}, \pm\frac{1}{12} \right\}$$

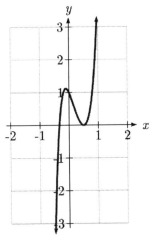

Testing $\dfrac{1}{2}$,

```
1/2 |  12   -20    19    -6    -2     1
     ↓        6    -7     6     0    -1
        12   -14    12     0    -2 |   0
```

Success! Because the graph bounces at this intercept, it is likely that this zero has multiplicity 2. We can try synthetic division again to test that.

```
1/2 |  12   -14    12     0    -2
     ↓        6    -4     4     2
        12    -8     8    -4 |   0
```

The other real root appears to be $-\dfrac{1}{3}$ or $-\dfrac{1}{4}$. Testing $-\dfrac{1}{3}$,

```
-1/3 |  12    -8     8    -4
      ↓       -4     4    -4
         12   -12    12 |   0
```

Excellent! So far, we have factored the polynomial to

$$f(x) = \left(x - \frac{1}{2}\right)^2 \left(x + \frac{1}{3}\right)\left(12x^2 - 12x + 12\right) = 12\left(x - \frac{1}{2}\right)^2 \left(x + \frac{1}{3}\right)\left(x^2 - x + 1\right)$$

We can use the quadratic formula to find the two remaining zeros by setting $x^2 - x + 1 = 0$, which are likely complex zeros.

$$x = \frac{1 \pm \sqrt{(-1)^2 - 4(1)(1)}}{2(1)} = \frac{1 \pm \sqrt{-3}}{2} = \frac{1 \pm i\sqrt{3}}{2}.$$

The zeros of the function are $x = \frac{1}{2}, -\frac{1}{3}, \frac{1 + i\sqrt{3}}{2}, \frac{1 - i\sqrt{3}}{2}$. We could write the function

fully factored as $f(x) = 12\left(x - \frac{1}{2}\right)^2 \left(x + \frac{1}{3}\right)\left(x - \frac{1 + i\sqrt{3}}{2}\right)\left(x - \frac{1 - i\sqrt{3}}{2}\right)$.

When factoring a polynomial like we did at the end of the last example, we say that it is **factored completely over the complex numbers**, meaning it is impossible to factor the polynomial any further using complex numbers. If we wanted to factor the function over

the **real numbers**, we would have stopped at $f(x) = 12\left(x - \frac{1}{2}\right)^2 \left(x + \frac{1}{3}\right)\left(x^2 - x + 1\right)$. Since

the zeros of $x^2 - x + 1$ are nonreal, we call $x^2 - x + 1$ an **irreducible quadratic** meaning it is impossible to break it down any further using real numbers.

It turns out that a polynomial with real number coefficients can be factored into a product of linear factors corresponding to the real zeros of the function and irreducible quadratic factors which give the nonreal zeros of the function. Consequently, any nonreal zeros will come in conjugate pairs, so if z is a zero of the polynomial, so is $\bar{z}$.

Try it Now

4. Find the real and complex zeros of $f(x) = x^3 - 4x^2 + 9x - 10$.

Important Topics of This Section

Complex and Imaginary numbers

Finding Complex zeros of polynomials

1. $(3-4i)-(2+5i)=1-9i$

2. $(3-4i)(2+3i)=18+i$

3. $x = \dfrac{-3 \pm \sqrt{(3)^2 - 4(2)(4)}}{2(2)} = \dfrac{-3 \pm \sqrt{-23}}{4} = \dfrac{-3 \pm i\sqrt{23}}{4} = \dfrac{-3}{4} \pm \dfrac{\sqrt{23}}{4}i$

4. Cauchy's Bound limits us to the interval [-11, 11]. The rational roots theorem gives a list of potential zeros: $\{\pm 1, \pm 2, \pm 5, \pm 10\}$. A quick graph shows that the likely rational root is $x = 2$.

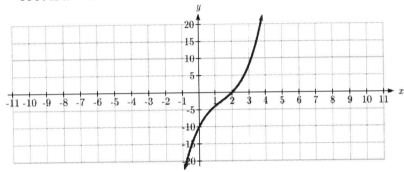

Verifying this,

$$
\begin{array}{r|rrrr}
2 & 1 & -4 & 9 & -10 \\
  &   & 2 & -4 & 10 \\
\hline
  & 1 & -2 & 5 & \boxed{0}
\end{array}
$$

So $f(x) = (x-2)(x^2 - 2x + 5)$

Using quadratic formula, we can find the complex roots from the irreducible quadratic.

$$x = \dfrac{-(-2) \pm \sqrt{(-2)^2 - 4(1)(5)}}{2(1)} = \dfrac{2 \pm \sqrt{-16}}{2} = \dfrac{-2 \pm 4i}{2} = -1 \pm 2i \ .$$

The zeros of this polynomial are $x = 2, -1+2i, -1-2i$

Section 3.6 Exercises

Simplify each expression to a single complex number.

1. $\sqrt{-9}$　　　　　　　2. $\sqrt{-16}$　　　　　　　3. $\sqrt{-6}\sqrt{-24}$

4. $\sqrt{-3}\sqrt{-75}$　　　　5. $\dfrac{2+\sqrt{-12}}{2}$　　　　6. $\dfrac{4+\sqrt{-20}}{2}$

Simplify each expression to a single complex number.

7. $(3+2i)+(5-3i)$　　　　　　　8. $(-2-4i)+(1+6i)$

9. $(-5+3i)-(6-i)$　　　　　　　10. $(2-3i)-(3+2i)$

11. $(2+3i)(4i)$　　　　　　　　12. $(5-2i)(3i)$

13. $(6-2i)(5)$　　　　　　　　14. $(-2+4i)(8)$

15. $(2+3i)(4-i)$　　　　　　　16. $(-1+2i)(-2+3i)$

17. $(4-2i)(4+2i)$　　　　　　　18. $(3+4i)(3-4i)$

19. $\dfrac{3+4i}{2}$　　　　　　　　20. $\dfrac{6-2i}{3}$

21. $\dfrac{-5+3i}{2i}$　　　　　　　22. $\dfrac{6+4i}{i}$

23. $\dfrac{2-3i}{4+3i}$　　　　　　　24. $\dfrac{3+4i}{2-i}$

Find all of the zeros of the polynomial then completely factor it over the real numbers and completely factor it over the complex numbers.

25. $f(x)=x^2-4x+13$　　　　　　26. $f(x)=x^2-2x+5$

27. $f(x)=3x^2+2x+10$　　　　　28. $f(x)=x^3-2x^2+9x-18$

29. $f(x)=x^3+6x^2+6x+5$　　　30. $f(x)=3x^3-13x^2+43x-13$

31. $f(x)=x^3+3x^2+4x+12$　　　32. $f(x)=4x^3-6x^2-8x+15$

33. $f(x)=x^3+7x^2+9x-2$　　　　34. $f(x)=9x^3+2x+1$

35. $f(x)=4x^4-4x^3+13x^2-12x+3$　　36. $f(x)=2x^4-7x^3+14x^2-15x+6$

37. $f(x)=x^4+x^3+7x^2+9x-18$　　38. $f(x)=6x^4+17x^3-55x^2+16x+12$

39. $f(x)=-3x^4-8x^3-12x^2-12x-5$　40. $f(x)=8x^4+50x^3+43x^2+2x-4$

41. $f(x)=x^4+9x^2+20$　　　　　42. $f(x)=x^4+5x^2-24$

Section 3.7 Rational Functions

In the previous sections, we have built polynomials based on the positive whole number power functions. In this section, we explore functions based on power functions with negative integer powers, called rational functions.

Example 1

You plan to drive 100 miles. Find a formula for the time the trip will take as a function of the speed you drive.

You may recall that multiplying speed by time will give you distance. If we let t represent the drive time in hours, and v represent the velocity (speed or rate) at which we drive, then $vt = \text{distance}$. Since our distance is fixed at 100 miles, $vt = 100$. Solving this relationship for the time gives us the function we desired:

$$t(v) = \frac{100}{v} = 100v^{-1}$$

While this type of relationship can be written using the negative exponent, it is more common to see it written as a fraction.

This particular example is one of an **inversely proportional** relationship – where one quantity is a constant divided by the other quantity, like $y = \frac{5}{x}$.

Notice that this is a transformation of the reciprocal toolkit function, $f(x) = \frac{1}{x}$

Several natural phenomena, such as gravitational force and volume of sound, behave in a manner **inversely proportional to the square** of another quantity. For example, the volume, V, of a sound heard at a distance d from the source would be related by $V = \frac{k}{d^2}$ for some constant value k.

These functions are transformations of the reciprocal squared toolkit function $f(x) = \frac{1}{x^2}$.

We have seen the graphs of the basic reciprocal function and the squared reciprocal function from our study of toolkit functions. These graphs have several important features.

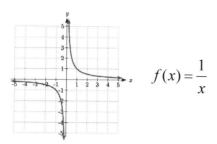 $f(x) = \dfrac{1}{x}$

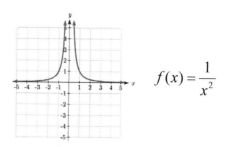 $f(x) = \dfrac{1}{x^2}$

Let's begin by looking at the reciprocal function, $f(x) = \dfrac{1}{x}$. As you well know, dividing by zero is not allowed and therefore zero is not in the domain, and so the function is undefined at an input of zero.

Short run behavior:
As the input values approach zero from the left side (taking on very small, negative values), the function values become very large in the negative direction (in other words, they approach negative infinity).
We write: as $x \to 0^-$, $f(x) \to -\infty$.

As we approach zero from the right side (small, positive input values), the function values become very large in the positive direction (approaching infinity).
We write: as $x \to 0^+$, $f(x) \to \infty$.

This behavior creates a **vertical asymptote**. An asymptote is a line that the graph approaches. In this case the graph is approaching the vertical line $x = 0$ as the input becomes close to zero.

Long run behavior:
As the values of x approach infinity, the function values approach 0.
As the values of x approach negative infinity, the function values approach 0.
Symbolically: as $x \to \pm\infty$, $f(x) \to 0$

Based on this long run behavior and the graph we can see that the function approaches 0 but never actually reaches 0, it just "levels off" as the inputs become large. This behavior creates a **horizontal asymptote**. In this case the graph is approaching the horizontal line $f(x) = 0$ as the input becomes very large in the negative and positive directions.

Vertical and Horizontal Asymptotes

A **vertical asymptote** of a graph is a vertical line $x = a$ where the graph tends towards positive or negative infinity as the inputs approach a. As $x \to a$, $f(x) \to \pm\infty$.

A **horizontal asymptote** of a graph is a horizontal line $y = b$ where the graph approaches the line as the inputs get large. As $x \to \pm\infty$, $f(x) \to b$.

1. Use symbolic notation to describe the long run behavior and
   short run behavior for the reciprocal squared function.

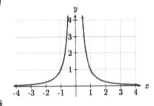

Example 2

Sketch a graph of the reciprocal function shifted two units to the left and up three units.
Identify the horizontal and vertical asymptotes of the graph, if any.

Transforming the graph left 2 and up 3 would result in the function

$f(x) = \dfrac{1}{x+2} + 3$, or equivalently, by giving the terms a common denominator,

$f(x) = \dfrac{3x+7}{x+2}$.

Shifting the toolkit function would give us
this graph. Notice that this equation is
undefined at $x = -2$, and the graph also is
showing a vertical asymptote at $x = -2$.

As $x \to -2^-$, $f(x) \to -\infty$, and

as $x \to -2^+$, $f(x) \to \infty$

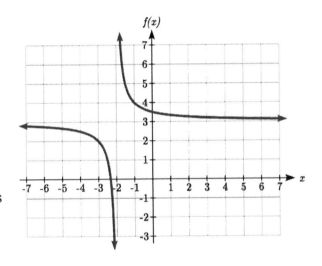

As the inputs grow large, the graph appears
to be leveling off at output values of 3,
indicating a horizontal asymptote at $y = 3$.

As $x \to \pm\infty$, $f(x) \to 3$.

Notice that horizontal and vertical asymptotes get shifted left 2 and up 3 along with the
function.

2. Sketch the graph and find the horizontal and vertical asymptotes of the reciprocal
   squared function that has been shifted right 3 units and down 4 units.

In the previous example, we shifted a toolkit function in a way that resulted in a function

of the form $f(x) = \dfrac{3x+7}{x+2}$. This is an example of a more general rational function.

Rational Function

A **rational function** is a function that can be written as the ratio of two polynomials, $P(x)$ and $Q(x)$.

$$f(x) = \frac{P(x)}{Q(x)} = \frac{a_0 + a_1 x + a_2 x^2 + \cdots + a_p x^p}{b_0 + b_1 x + b_2 x^2 + \cdots + b_q x^q}$$

Example 3

A large mixing tank currently contains 100 gallons of water, into which 5 pounds of sugar have been mixed. A tap will open pouring 10 gallons per minute of water into the tank at the same time sugar is poured into the tank at a rate of 1 pound per minute. Find the concentration (pounds per gallon) of sugar in the tank after t minutes.

Notice that the amount of water in the tank is changing linearly, as is the amount of sugar in the tank. We can write an equation independently for each:
$water = 100 + 10t$
$sugar = 5 + 1t$

The concentration, C, will be the ratio of pounds of sugar to gallons of water

$$C(t) = \frac{5 + t}{100 + 10t}$$

Finding Asymptotes and Intercepts

Given a rational function, as part of investigating the short run behavior we are interested in finding any vertical and horizontal asymptotes, as well as finding any vertical or horizontal intercepts, as we have done in the past.

To find vertical asymptotes, we notice that the vertical asymptotes in our examples occur when the denominator of the function is undefined. With one exception, a vertical asymptote will occur whenever the denominator is undefined.

Example 4

Find the vertical asymptotes of the function $k(x) = \dfrac{5 + 2x^2}{2 - x - x^2}$

To find the vertical asymptotes, we determine where this function will be undefined by setting the denominator equal to zero:
$2 - x - x^2 = 0$
$(2 + x)(1 - x) = 0$
$x = -2, 1$

This indicates two vertical asymptotes, which a look at a graph confirms.

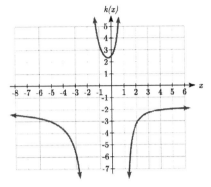

The exception to this rule can occur when both the numerator and denominator of a rational function are zero at the same input.

Example 5

Find the vertical asymptotes of the function $k(x) = \dfrac{x-2}{x^2-4}$.

To find the vertical asymptotes, we determine where this function will be undefined by setting the denominator equal to zero:

$x^2 - 4 = 0$

$x^2 = 4$

$x = -2, 2$

However, the numerator of this function is also equal to zero when $x = 2$. Because of this, the function will still be undefined at 2, since $\dfrac{0}{0}$ is undefined, but the graph will not have a vertical asymptote at $x = 2$.

The graph of this function will have the vertical asymptote at $x = -2$, but at $x = 2$ the graph will have a hole: a single point where the graph is not defined, indicated by an open circle.

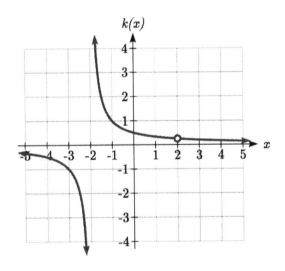

Vertical Asymptotes and Holes of Rational Functions

The **vertical asymptotes** of a rational function will occur where the denominator of the function is equal to zero and the numerator is not zero.

A **hole** occurs in the graph of a rational function if an input causes both numerator and denominator to be zero. In this case, factor the numerator and denominator and simplify; if the simplified expression still has a zero in the denominator at the original input the original function has a vertical asymptote at the input, otherwise it has a hole.

To find horizontal asymptotes, we are interested in the behavior of the function as the input grows large, so we consider long run behavior of the numerator and denominator separately. Recall that a polynomial's long run behavior will mirror that of the leading term. Likewise, a rational function's long run behavior will mirror that of the ratio of the leading terms of the numerator and denominator functions.

There are three distinct outcomes when this analysis is done:

Case 1: The degree of the denominator > degree of the numerator

Example: $f(x) = \dfrac{3x+2}{x^2+4x-5}$

In this case, the long run behavior is $f(x) \approx \dfrac{3x}{x^2} = \dfrac{3}{x}$. This tells us that as the inputs grow large, this function will behave similarly to the function $g(x) = \dfrac{3}{x}$. As the inputs grow large, the outputs will approach zero, resulting in a horizontal asymptote at $y = 0$.
As $x \to \pm\infty$, $f(x) \to 0$

Case 2: The degree of the denominator < degree of the numerator

Example: $f(x) = \dfrac{3x^2+2}{x-5}$

In this case, the long run behavior is $f(x) \approx \dfrac{3x^2}{x} = 3x$. This tells us that as the inputs grow large, this function will behave similarly to the function $g(x) = 3x$. As the inputs grow large, the outputs will grow and not level off, so this graph has no horizontal asymptote.
As $x \to \pm\infty$, $f(x) \to \pm\infty$, respectively.

Case 3: The degree of the denominator = degree of the numerator

Example: $f(x) = \dfrac{3x^2+2}{x^2+4x-5}$

In this case, the long run behavior is $f(x) \approx \dfrac{3x^2}{x^2} = 3$. This tells us that as the inputs grow large, this function will behave like the function $g(x) = 3$, which is a horizontal line. As $x \to \pm\infty$, $f(x) \to 3$, resulting in a horizontal asymptote at $y = 3$.

> ### Horizontal Asymptote of Rational Functions
>
> The **horizontal asymptote** of a rational function can be determined by looking at the degrees of the numerator and denominator.
>
> Degree of denominator > degree of numerator: Horizontal asymptote at $y = 0$
>
> Degree of denominator < degree of numerator: No horizontal asymptote
>
> Degree of denominator = degree of numerator: Horizontal asymptote at ratio of leading coefficients.

Example 6

In the sugar concentration problem from earlier, we created the equation

$$C(t) = \frac{5+t}{100+10t}.$$

Find the horizontal asymptote and interpret it in context of the scenario.

Both the numerator and denominator are linear (degree 1), so since the degrees are equal, there will be a horizontal asymptote at the ratio of the leading coefficients. In the numerator, the leading term is t, with coefficient 1. In the denominator, the leading term is $10t$, with coefficient 10. The horizontal asymptote will be at the ratio of these values: As $t \to \infty$, $C(t) \to \frac{1}{10}$. This function will have a horizontal asymptote at

$$y = \frac{1}{10}.$$

This tells us that as the input gets large, the output values will approach 1/10. In context, this means that as more time goes by, the concentration of sugar in the tank will approach one tenth of a pound of sugar per gallon of water or 1/10 pounds per gallon.

Example 7

Find the horizontal and vertical asymptotes of the function

$$f(x) = \frac{(x-2)(x+3)}{(x-1)(x+2)(x-5)}$$

First, note this function has no inputs that make both the numerator and denominator zero, so there are no potential holes. The function will have vertical asymptotes when the denominator is zero, causing the function to be undefined. The denominator will be zero at $x = 1$, -2, and 5, indicating vertical asymptotes at these values.

The numerator has degree 2, while the denominator has degree 3. Since the degree of the denominator is greater than that of the numerator, the denominator will grow faster than the numerator, causing the outputs to tend towards zero as the inputs get large, and so as $x \to \pm\infty$, $f(x) \to 0$. This function will have a horizontal asymptote at $y = 0$.

Try it Now

3. Find the vertical and horizontal asymptotes of the function $f(x) = \dfrac{(2x-1)(2x+1)}{(x-2)(x+3)}$

Intercepts

As with all functions, a rational function will have a vertical intercept when the input is zero, if the function is defined at zero. It is possible for a rational function to not have a vertical intercept if the function is undefined at zero.

Likewise, a rational function will have horizontal intercepts at the inputs that cause the output to be zero (unless that input corresponds to a hole). It is possible there are no horizontal intercepts. Since a fraction is only equal to zero when the numerator is zero, horizontal intercepts will occur when the numerator of the rational function is equal to zero.

Example 8

Find the intercepts of $f(x) = \dfrac{(x-2)(x+3)}{(x-1)(x+2)(x-5)}$

We can find the vertical intercept by evaluating the function at zero
$$f(0) = \frac{(0-2)(0+3)}{(0-1)(0+2)(0-5)} = \frac{-6}{10} = -\frac{3}{5}$$

The horizontal intercepts will occur when the function is equal to zero:
$$0 = \frac{(x-2)(x+3)}{(x-1)(x+2)(x-5)}$$ This is zero when the numerator is zero
$$0 = (x-2)(x+3)$$
$$x = 2, -3$$

Try it Now

4. Given the reciprocal squared function that is shifted right 3 units and down 4 units, write this as a rational function and find the horizontal and vertical intercepts and the horizontal and vertical asymptotes.

From the previous example, you probably noticed that the numerator of a rational function reveals the horizontal intercepts of the graph, while the denominator reveals the vertical asymptotes of the graph. As with polynomials, factors of the numerator may have integer powers greater than one. Happily, the effect on the shape of the graph at those intercepts is the same as we saw with polynomials.

When factors of the denominator have integer powers greater than one, the behavior at the corresponding vertical asymptote will mirror one of the two toolkit reciprocal functions.

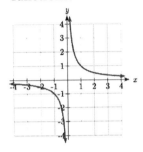

We get this behavior when the degree of the factor in the denominator is odd. The distinguishing characteristic is that on one side of the vertical asymptote the graph heads towards positive infinity, and on the other side the graph heads towards negative infinity.

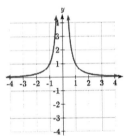

We get this behavior when the degree of the factor in the denominator is even. The distinguishing characteristic is that the graph either heads toward positive infinity on both sides of the vertical asymptote, or heads toward negative infinity on both sides.

For example, the graph of

$$f(x) = \frac{(x+1)^2(x-3)}{(x+3)^2(x-2)}$$ is shown here.

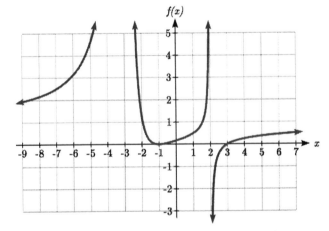

At the horizontal intercept $x = -1$ corresponding to the $(x+1)^2$ factor of the numerator, the graph bounces at the intercept, consistent with the quadratic nature of the factor.

At the horizontal intercept $x = 3$ corresponding to the $(x-3)$ factor of the numerator, the graph passes through the axis as we'd expect from a linear factor.

At the vertical asymptote $x = -3$ corresponding to the $(x+3)^2$ factor of the denominator, the graph heads towards positive infinity on both sides of the asymptote, consistent with the behavior of the $\dfrac{1}{x^2}$ toolkit.

At the vertical asymptote $x = 2$ corresponding to the $(x - 2)$ factor of the denominator, the graph heads towards positive infinity on the left side of the asymptote and towards negative infinity on the right side, consistent with the behavior of the $\dfrac{1}{x}$ toolkit.

Example 9

Sketch a graph of $f(x) = \dfrac{(x+2)(x-3)}{(x+1)^2(x-2)}$.

We can start our sketch by finding intercepts and asymptotes. Evaluating the function at zero gives the vertical intercept:

$$f(0) = \frac{(0+2)(0-3)}{(0+1)^2(0-2)} = 3$$

Looking at when the numerator of the function is zero, we can determine the graph will have horizontal intercepts at $x = -2$ and $x = 3$. At each, the behavior will be linear, with the graph passing through the intercept.

Looking at when the denominator of the function is zero, we can determine the graph will have vertical asymptotes at $x = -1$ and $x = 2$.

Finally, the degree of denominator is larger than the degree of the numerator, telling us this graph has a horizontal asymptote at $y = 0$.

To sketch the graph, we might start by plotting the three intercepts. Since the graph has no horizontal intercepts between the vertical asymptotes, and the vertical intercept is positive, we know the function must remain positive between the asymptotes, letting us fill in the middle portion of the graph.

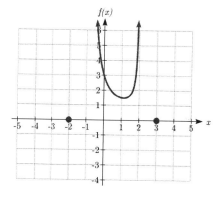

Since the factor associated with the vertical asymptote at $x = -1$ was squared, we know the graph will have the same behavior on both sides of the asymptote. Since the graph heads towards positive infinity as the inputs approach the asymptote on the right, the graph will head towards positive infinity on the left as well. For the vertical asymptote at $x = 2$, the factor was not squared, so the graph will have opposite behavior on either side of the asymptote.

After passing through the horizontal intercepts, the graph will then level off towards an output of zero, as indicated by the horizontal asymptote.

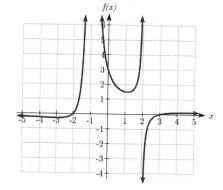

5. Given the function $f(x) = \dfrac{(x+2)^2(x-2)}{2(x-1)^2(x-3)}$, use the characteristics of polynomials and rational functions to describe its behavior and sketch the function.

Since a rational function written in factored form will have a horizontal intercept where each factor of the numerator is equal to zero, we can form a numerator that will pass through a set of horizontal intercepts by introducing a corresponding set of factors. Likewise, since the function will have a vertical asymptote where each factor of the denominator is equal to zero, we can form a denominator that will produce the vertical asymptotes by introducing a corresponding set of factors.

Writing Rational Functions from Intercepts and Asymptotes

If a rational function has horizontal intercepts at $x = x_1, x_2, \ldots, x_n$, and vertical asymptotes at $x = v_1, v_2, \ldots, v_m$ then the function can be written in the form

$$f(x) = a\frac{(x-x_1)^{P_1}(x-x_2)^{P_2}\cdots(x-x_n)^{P_n}}{(x-v_1)^{q_1}(x-v_2)^{q_2}\cdots(x-v_m)^{q_n}}$$

where the powers p_i or q_i on each factor can be determined by the behavior of the graph at the corresponding intercept or asymptote, and the stretch factor a can be determined given a value of the function other than the horizontal intercept, or by the horizontal asymptote if it is nonzero.

Example 10

Write an equation for the rational function graphed here.

The graph appears to have horizontal intercepts at $x = -2$ and $x = 3$. At both, the graph passes through the intercept, suggesting linear factors.

The graph has two vertical asymptotes. The one at $x = -1$ seems to exhibit the basic behavior similar to $\dfrac{1}{x}$, with the graph heading toward positive infinity on one side and heading toward negative infinity on the other.

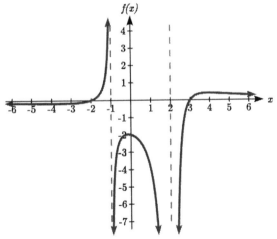

The asymptote at $x = 2$ is exhibiting a behavior similar to $\dfrac{1}{x^2}$, with the graph heading toward negative infinity on both sides of the asymptote.

Utilizing this information indicates an function of the form

$$f(x) = a\frac{(x+2)(x-3)}{(x+1)(x-2)^2}$$

To find the stretch factor, we can use another clear point on the graph, such as the vertical intercept (0,-2):

$$-2 = a\frac{(0+2)(0-3)}{(0+1)(0-2)^2}$$

$$-2 = a\frac{-6}{4}$$

$$a = \frac{-8}{-6} = \frac{4}{3}$$

This gives us a final function of $f(x) = \dfrac{4(x+2)(x-3)}{3(x+1)(x-2)^2}$

Oblique Asymptotes

Earlier we saw graphs of rational functions that had no horizontal asymptote, which occurs when the degree of the numerator is larger than the degree of the denominator. We can, however, describe in more detail the long-run behavior of a rational function.

Example 11

Describe the long-run behavior of $f(x) = \dfrac{3x^2 + 2}{x - 5}$

Earlier we explored this function when discussing horizontal asymptotes. We found the long-run behavior is $f(x) \approx \dfrac{3x^2}{x} = 3x$, meaning that $x \to \pm\infty$, $f(x) \to \pm\infty$, respectively, and there is no horizontal asymptote.

If we were to do polynomial long division, we could get a better understanding of the behavior as $x \to \pm\infty$.

$$\begin{array}{r} 3x+15 \\ x-5{\overline{\smash{\big)}\,3x^2+0x+2}} \\ \underline{-\left(3x^2-15x\right)} \\ 15x+2 \\ \underline{-\left(15x-75\right)} \\ 77 \end{array}$$

This means $f(x)=\dfrac{3x^2+2}{x-5}$ can be rewritten as

$f(x)=3x+15+\dfrac{77}{x-5}$.

As $x\to\pm\infty$, the term $\dfrac{77}{x-5}$ will become very

small and approach zero, becoming insignificant. The remaining $3x+15$ then describes the long-run behavior of the function: as $x\to\pm\infty$, $f(x)\to 3x+15$.

We call this equation $y=3x+15$ the **oblique asymptote** of the function.

In the graph, you can see how the function is approaching the line on the far left and far right.

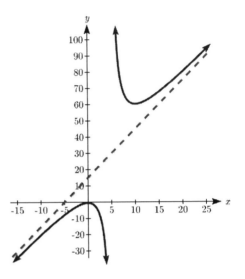

Oblique Asymptotes

To explore the long-run behavior of a rational function,

1) Perform polynomial long division (or synthetic division)

2) The quotient will describe the asymptotic behavior of the function

When this result is a line, we call it an **oblique asymptote**, or **slant asymptote**.

Example 12

Find the oblique asymptote of $f(x)=\dfrac{-x^2+2x+1}{x+1}$

Performing polynomial long division:

$$x+1{\overline{\smash{\big)}\,-x^2+2x+1}}$$ with quotient $-x+3$ at top

$$\begin{array}{r} -x+3 \\ x+1{\overline{\smash{\big)}\,-x^2+2x+1}} \\ \underline{-\left(-x^2-x\right)} \\ 3x+1 \\ \underline{-\left(3x+3\right)} \\ -2 \end{array}$$

This allows us to rewrite the function as

$$f(x) = -x+3-\frac{2}{x+1}.$$

The quotient, $y = -x+3$, is the oblique asymptote of $f(x)$. Just like functions we saw earlier approached their horizontal asymptote in the long run, this function will approach this oblique asymptote in the long run.

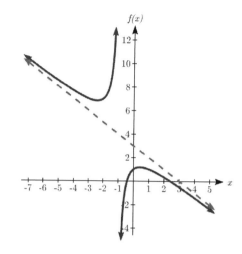

Try it Now

6. Find the oblique asymptote of $f(x) = \dfrac{1+7x-2x^2}{x-2}$

While we primarily concern ourselves with oblique asymptotes, this same approach can describe other asymptotic behavior.

Example 13

Describe the long-run shape of $f(x) = \dfrac{-x^3-x^2+4x+2}{x+1}$

We could rewrite this using long division as

$$f(x) = -x^2+4+\frac{2}{x+1}.$$

Just looking at the quotient gives us the asymptote, $y = -x^2+4$.

This suggests that in the long run, the function will behave like a downwards opening parabola. The function will also have a vertical asymptote at $x = -1$.

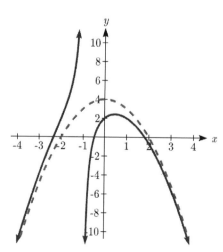

Important Topics of this Section
Inversely proportional; Reciprocal toolkit function
Inversely proportional to the square; Reciprocal squared toolkit function
Horizontal Asymptotes
Vertical Asymptotes
Rational Functions
Finding intercepts, asymptotes, and holes.
Given equation sketch the graph
Identifying a function from its graph
Oblique Asymptotes

Try it Now Answers

1. Long run behavior, as $x \to \pm\infty$, $f(x) \to 0$

   Short run behavior, as $x \to 0$, $f(x) \to \infty$ (there are no horizontal or vertical intercepts)

2. The function and the asymptotes are shifted 3 units right and 4 units down.
   As $x \to 3$, $f(x) \to \infty$ and as $x \to \pm\infty$, $f(x) \to -4$

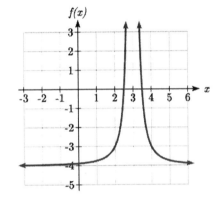

3. Vertical asymptotes at $x = 2$ and $x = -3$; horizontal asymptote at $y = 4$

4. For the transformed reciprocal squared function, we find the rational form.

$$f(x) = \frac{1}{(x-3)^2} - 4 = \frac{1-4(x-3)^2}{(x-3)^2} = \frac{1-4(x^2-6x+9)}{(x-3)(x-3)} = \frac{-4x^2+24x-35}{x^2-6x+9}$$

Since the numerator is the same degree as the denominator we know that as $x \to \pm\infty$, $f(x) \to -4$. $y = -4$ is the horizontal asymptote. Next, we set the denominator equal to zero to find the vertical asymptote at $x = 3$, because as $x \to 3$, $f(x) \to \infty$. We set the numerator equal to 0 and find the horizontal intercepts are at (2.5,0) and (3.5,0), then we evaluate at 0 and the vertical intercept is at $\left(0, \dfrac{-35}{9}\right)$

Try it Now Answers, Continued

5. Horizontal asymptote at $y = 1/2$.
    Vertical asymptotes are at $x = 1$, and $x = 3$.
    Vertical intercept at (0, 4/3),
    Horizontal intercepts (2, 0) and (-2, 0)
    (-2, 0) is a double zero and the graph bounces off the
    axis at this point.
    (2, 0) is a single zero and crosses the axis at this point.

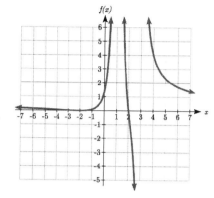

6. Using long division:

$$
\begin{array}{r}
-2x+3 \\
x-2 \overline{\smash{\big)}\ -2x^2+7x+1} \\
\underline{-\left(-2x^2+4x\right)} \\
3x+1 \\
\underline{-\left(3x-6\right)} \\
7
\end{array}
$$

$$f(x) = \frac{1+7x-2x^2}{x-2} = -2x+3+\frac{7}{x-2}$$

The oblique asymptote is $y = -2x+3$

Section 3.7 Exercises

Match each equation form with one of the graphs.

1. $f(x) = \dfrac{x-A}{x-B}$ 2. $g(x) = \dfrac{(x-A)^2}{x-B}$ 3. $h(x) = \dfrac{x-A}{(x-B)^2}$ 4. $k(x) = \dfrac{(x-A)^2}{(x-B)^2}$

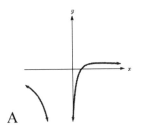 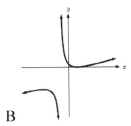 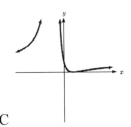 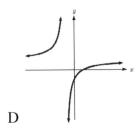

A   B   C   D

For each function, find the horizontal intercepts, the vertical intercept, the vertical asymptotes, and the horizontal asymptote. Use that information to sketch a graph.

5. $p(x) = \dfrac{2x-3}{x+4}$

6. $q(x) = \dfrac{x-5}{3x-1}$

7. $s(x) = \dfrac{4}{(x-2)^2}$

8. $r(x) = \dfrac{5}{(x+1)^2}$

9. $f(x) = \dfrac{3x^2 - 14x - 5}{3x^2 + 8x - 16}$

10. $g(x) = \dfrac{2x^2 + 7x - 15}{3x^2 - 14 + 15}$

11. $a(x) = \dfrac{x^2 + 2x - 3}{x^2 - 1}$

12. $b(x) = \dfrac{x^2 - x - 6}{x^2 - 4}$

13. $h(x) = \dfrac{2x^2 + x - 1}{x-4}$

14. $k(x) = \dfrac{2x^2 - 3x - 20}{x-5}$

15. $n(x) = \dfrac{3x^2 + 4x - 4}{x^3 - 4x^2}$

16. $m(x) = \dfrac{5-x}{2x^2 + 7x + 3}$

17. $w(x) = \dfrac{(x-1)(x+3)(x-5)}{(x+2)^2 (x-4)}$

18. $z(x) = \dfrac{(x+2)^2 (x-5)}{(x-3)(x+1)(x+4)}$

Write an equation for a rational function with the given characteristics.

19. Vertical asymptotes at $x = 5$ and $x = -5$

    x intercepts at $(2,0)$ and $(-1,0)$ y intercept at $(0,4)$

20. Vertical asymptotes at $x = -4$ and $x = -1$

    x intercepts at $(1,0)$ and $(5,0)$ y intercept at $(0,7)$

21. Vertical asymptotes at $x = -4$ and $x = -5$

    x intercepts at $(4,0)$ and $(-6,0)$ Horizontal asymptote at $y = 7$

22. Vertical asymptotes at $x = -3$ and $x = 6$

    x intercepts at $(-2,0)$ and $(1,0)$ Horizontal asymptote at $y = -2$

23. Vertical asymptote at $x = -1$
    Double zero at $x = 2$ y intercept at $(0,2)$

24. Vertical asymptote at $x = 3$
    Double zero at $x = 1$ y intercept at $(0,4)$

Write an equation for the function graphed.

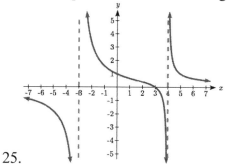

25.

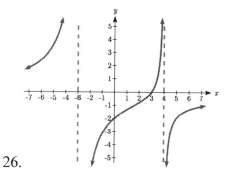

26.

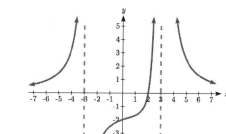

27.

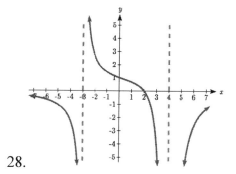

28.

Write an equation for the function graphed.

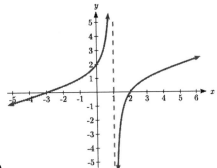

29.

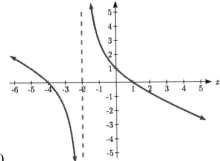

30.

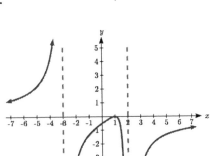

31.

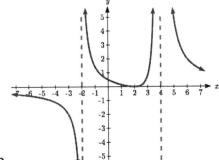

32.

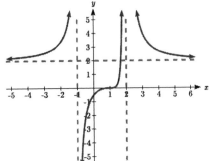

33.

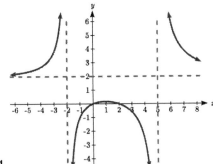

34.

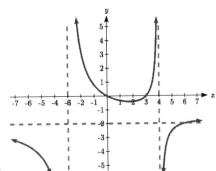

35.

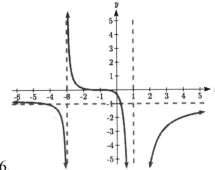

36.

Write an equation for the function graphed.

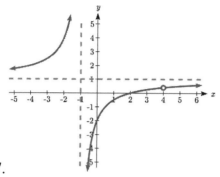

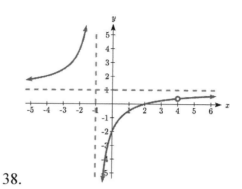

37.

38.

Find the oblique asymptote of each function.

39. $f(x) = \dfrac{3x^2 + 4x}{x + 2}$

40. $g(x) = \dfrac{2x^2 + 3x - 8}{x - 1}$

41. $h(x) = \dfrac{x^2 - x - 3}{2x - 6}$

42. $k(x) = \dfrac{5 + x - 2x^2}{2x + 1}$

43. $m(x) = \dfrac{-2x^3 + x^2 - 6x + 7}{x^2 + 3}$

44. $n(x) = \dfrac{2x^3 + x^2 + x}{x^2 + x + 1}$

45. A scientist has a beaker containing 20 mL of a solution containing 20% acid. To dilute this, she adds pure water.
   a. Write an equation for the concentration in the beaker after adding n mL of water.
   b. Find the concentration if 10 mL of water has been added.
   c. How many mL of water must be added to obtain a 4% solution?
   d. What is the behavior as $n \to \infty$, and what is the physical significance of this?

46. A scientist has a beaker containing 30 mL of a solution containing 3 grams of potassium hydroxide. To this, she mixes a solution containing 8 milligrams per mL of potassium hydroxide.
   a. Write an equation for the concentration in the tank after adding n mL of the second solution.
   b. Find the concentration if 10 mL of the second solution has been added.
   c. How many mL of water must be added to obtain a 50 mg/mL solution?
   d. What is the behavior as $n \to \infty$, and what is the physical significance of this?

47. Oscar is hunting magnetic fields with his gauss meter, a device for measuring the strength and polarity of magnetic fields. The reading on the meter will increase as Oscar gets closer to a magnet. Oscar is in a long hallway at the end of which is a room containing an extremely strong magnet. When he is far down the hallway from the room, the meter reads a level of 0.2. He then walks down the hallway and enters the room. When he has gone 6 feet into the room, the meter reads 2.3. Eight feet into the room, the meter reads 4.4. [UW]

    a. Give a rational model of form $m(x) = \dfrac{ax+b}{cx+d}$ relating the meter reading $m(x)$ to how many feet x Oscar has gone into the room.

    b. How far must he go for the meter to reach 10? 100?

    c. Considering your function from part (a) and the results of part (b), how far into the room do you think the magnet is?

48. The more you study for a certain exam, the better your performance on it. If you study for 10 hours, your score will be 65%. If you study for 20 hours, your score will be 95%. You can get as close as you want to a perfect score just by studying long enough. Assume your percentage score, $p(n)$, is a function of the number of hours, n, that you study in the form $p(n) = \dfrac{an+b}{cn+d}$. If you want a score of 80%, how long do you need to study? [UW]

49. A street light is 10 feet north of a straight bike path that runs east-west. Olav is bicycling down the path at a rate of 15 miles per hour. At noon, Olav is 33 feet west of the point on the bike path closest to the street light. (See the picture). The relationship between the intensity C of light (in candlepower) and the

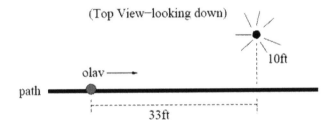

distance d (in feet) from the light source is given by $C = \dfrac{k}{d^2}$, where k is a constant depending on the light source. [UW]

    a. From 20 feet away, the street light has an intensity of 1 candle. What is k?

    b. Find a function which gives the intensity of the light shining on Olav as a function of time, in seconds.

    c. When will the light on Olav have maximum intensity?

    d. When will the intensity of the light be 2 candles?

Section 3.8 Inverses and Radical Functions

In this section, we will explore the inverses of polynomial and rational functions, and in particular the radical functions that arise in the process.

Example 1

A water runoff collector is built in the shape of a parabolic trough as shown below. Find the surface area of the water in the trough as a function of the depth of the water.

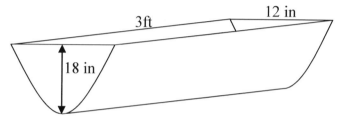

Since it will be helpful to have an equation for the parabolic cross-sectional shape, we will impose a coordinate system at the cross section, with x measured horizontally and y measured vertically, with the origin at the vertex of the parabola.

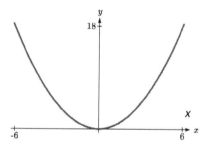

From this we find an equation for the parabolic shape. Since we placed the origin at the vertex of the parabola, we know the equation will have form $y(x) = ax^2$. Our equation will need to pass through the point (6,18), from which we can solve for the stretch factor a:

$$18 = a6^2$$

$$a = \frac{18}{36} = \frac{1}{2}$$

Our parabolic cross section has equation $y(x) = \frac{1}{2}x^2$

Since we are interested in the surface area of the water, we are interested in determining the width at the top of the water as a function of the water depth. For any depth y the width will be given by $2x$, so we need to solve the equation above for x. However notice that the original function is not one-to-one, and indeed given any output there are two inputs that produce the same output, one positive and one negative.

To find an inverse, we can restrict our original function to a limited domain on which it *is* one-to-one. In this case, it makes sense to restrict ourselves to positive *x* values. On this domain, we can find an inverse by solving for the input variable:

$$y = \frac{1}{2}x^2$$

$$2y = x^2$$

$$x = \pm\sqrt{2y}$$

This is not a function as written. Since we are limiting ourselves to positive *x* values, we eliminate the negative solution, giving us the inverse function we're looking for

$$x(y) = \sqrt{2y}$$

Since *x* measures from the center out, the entire width of the water at the top will be 2*x*. Since the trough is 3 feet (36 inches) long, the surface area will then be 36(2*x*), or in terms of *y*:

$$Area = 72x = 72\sqrt{2y}$$

The previous example illustrated two important things:
1) When finding the inverse of a quadratic, we have to limit ourselves to a domain on which the function is one-to-one.
2) The inverse of a quadratic function is a square root function. Both are toolkit functions and different types of power functions.

Functions involving roots are often called **radical functions**.

Example 2

Find the inverse of $f(x) = (x-2)^2 - 3 = x^2 - 4x + 1$

From the transformation form of the function, we can see this is a transformed quadratic with vertex at (2,-3) that opens upwards. Since the graph will be decreasing on one side of the vertex, and increasing on the other side, we can restrict this function to a domain on which it will be one-to-one by limiting the domain to $x \geq 2$.

To find the inverse, we will use the vertex form of the quadratic. We start by replacing the *f(x)* with a simple variable *y*, then solve for *x*.

$$y = (x-2)^2 - 3 \qquad \text{Add 3 to both sides}$$

$$y + 3 = (x-2)^2 \qquad \text{Take the square root}$$

$$\pm\sqrt{y+3} = x - 2 \qquad \text{Add 2 to both sides}$$

$$2 \pm \sqrt{y+3} = x$$

Of course, as written this is not a function. Since we restricted our original function to a domain of $x \geq 2$, the outputs of the inverse should be the same, telling us to utilize the positive case:

$$x = f^{-1}(y) = 2 + \sqrt{y+3}$$

If the quadratic had not been given in vertex form, rewriting it into vertex form is probably the best approach. Alternatively, we could have taken the standard equation and rewritten it equal to zero:

$$0 = x^2 - 4x + 1 - y$$

We would then be able to use the quadratic formula with $a = 1$, $b = -4$, and $c = (1-y)$, resulting in the same solutions we found above:

$$x = \frac{-(-4) \pm \sqrt{(-4)^2 - 4(1)(1-y)}}{2} = 2 \pm \frac{\sqrt{12+4y}}{2} = 2 \pm \sqrt{3+y}$$

Try it Now

1. Find the inverse of the function $f(x) = x^2 + 1$, on the domain $x \geq 0$.

While it is not possible to find an inverse of most polynomial functions, some other basic polynomials are invertible.

Example 3

Find the inverse of the function $f(x) = 5x^3 + 1$.

This is a transformation of the basic cubic toolkit function, and based on our knowledge of that function, we know it is one-to-one. Solving for the inverse by solving for x

$$y = 5x^3 + 1$$

$$y - 1 = 5x^3$$

$$\frac{y-1}{5} = x^3$$

$$x = f^{-1}(y) = \sqrt[3]{\frac{y-1}{5}}$$

Notice that this inverse is also a transformation of a power function with a fractional power, $x^{1/3}$.

Try it Now
2. Which toolkit functions have inverse functions without restricting their domain?

Besides being important as an inverse function, radical functions are common in important physical models.

Example 4

The velocity, v in feet per second, of a car that slammed on its brakes can be determined based on the length of skid marks that the tires left on the ground. This relationship is given by

$$v(d) = \sqrt{2gfd}$$

In this formula, g represents acceleration due to gravity (32 ft/sec²), d is the length of the skid marks in feet, and f is a constant representing the friction of the surface. A car lost control on wet asphalt, with a friction coefficient of 0.5, leaving 200 foot skid marks. How fast was the car travelling when it lost control?

Using the given values of $f = 0.5$ and $d = 200$, we can evaluate the given formula:

$v(200) = \sqrt{2(32)(0.5)(200)} = 80$ ft/sec , which is about 54.5 miles per hour.

When radical functions are composed with other functions, determining domain can become more complicated.

Example 5

Find the domain of the function $f(x) = \sqrt{\dfrac{(x+2)(x-3)}{(x-1)}}$.

Since a square root is only defined when the quantity under the radical is non-negative, we need to determine where $\dfrac{(x+2)(x-3)}{(x-1)} \geq 0$. A rational function can change signs (change from positive to negative or vice versa) at horizontal intercepts and at vertical asymptotes. For this equation, the graph could change signs at $x = -2$, 1, and 3.

To determine on which intervals the rational expression is positive, we could evaluate the expression at test values, or sketch a graph. While both approaches work equally well, for this example we will use a graph.

This function has two horizontal intercepts, both of which exhibit linear behavior, where the graph will pass through the intercept. There is one vertical asymptote, corresponding to a linear factor, leading to a behavior similar to the basic reciprocal toolkit function. There is a vertical intercept at (0, 6). This graph does not have a horizontal asymptote, since the degree of the numerator is larger than the degree of the denominator.

From the vertical intercept and horizontal intercept at $x = -2$, we can sketch the left side of the graph. From the behavior at the asymptote, we can sketch the right side of the graph.

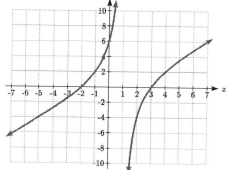

From the graph, we can now tell on which intervals this expression will be non-negative, so the original function $f(x)$ will be defined. $f(x)$ has domain $-2 \le x < 1$ *or* $x \ge 3$, or in interval notation, $[-2,1) \cup [3,\infty)$.

Like with finding inverses of quadratic functions, it is sometimes desirable to find the inverse of a rational function, particularly of rational functions that are the ratio of linear functions, such as our concentration examples.

Example 6

The function $C(n) = \dfrac{20 + 0.4n}{100 + n}$ was used in the previous section to represent the concentration of an acid solution after n mL of 40% solution has been added to 100 mL of a 20% solution. We might want to be able to determine instead how much 40% solution has been added based on the current concentration of the mixture.

To do this, we would want the inverse of this function:

$C = \dfrac{20 + 0.4n}{100 + n}$ multiply both sides by the denominator

$C(100 + n) = 20 + 0.4n$ distribute

$100C + Cn = 20 + 0.4n$ group everything with n on one side

$100C - 20 = 0.4n - Cn$ factor out n

$100C - 20 = (0.4 - C)n$ divide to find the inverse

$n(C) = \dfrac{100C - 20}{0.4 - C}$

If, for example, we wanted to know how many mL of 40% solution need to be added to obtain a concentration of 35%, we can simply evaluate the inverse rather than solving an equation involving the original function:

$n(0.35) = \dfrac{100(0.35) - 20}{0.4 - 0.35} = \dfrac{15}{0.05} = 300$ mL of 40% solution would need to be added.

Try it Now

3. Find the inverse of the function $f(x) = \dfrac{x+3}{x-2}$.

Important Topics of this Section

Imposing a coordinate system

Finding an inverse function

      Restricting the domain

Invertible toolkit functions

Radical Functions

Inverses of rational functions

Try it Now Answers

1. $y = x^2 + 1$

   $y - 1 = x^2$

   $x = f^{-1}(y) = \sqrt{y-1}$

2. identity, cubic, square root, cube root

3. $y = \dfrac{x+3}{x-2}$

   $y(x-2) = x + 3$

   $yx - 2y = x + 3$

   $yx - x = 2y + 3$

   $x(y-1) = 2y + 3$

   $f^{-1}(y) = \dfrac{2y+3}{y-1}$

Section 3.8 Exercises

For each function, find a domain on which the function is one-to-one and non-decreasing, then find an inverse of the function on this domain.

1. $f(x) = (x-4)^2$

2. $f(x) = (x+2)^2$

3. $f(x) = 12 - x^2$

4. $f(x) = 9 - x^2$

5. $f(x) = 3x^3 + 1$

6. $f(x) = 4 - 2x^3$

Find the inverse of each function.

7. $f(x) = 9 + \sqrt{4x-4}$

8. $f(x) = \sqrt{6x-8} + 5$

9. $f(x) = 9 + 2\sqrt[3]{x}$

10. $f(x) = 3 - \sqrt[3]{x}$

11. $f(x) = \dfrac{2}{x+8}$

12. $f(x) = \dfrac{3}{x-4}$

13. $f(x) = \dfrac{x+3}{x+7}$

14. $f(x) = \dfrac{x-2}{x+7}$

15. $f(x) = \dfrac{3x+4}{5-4x}$

16. $f(x) = \dfrac{5x+1}{2-5x}$

Police use the formula $v = \sqrt{20L}$ to estimate the speed of a car, v, in miles per hour, based on the length, L, in feet, of its skid marks when suddenly braking on a dry, asphalt road.

17. At the scene of an accident, a police officer measures a car's skid marks to be 215 feet long. Approximately how fast was the car traveling?

18. At the scene of an accident, a police officer measures a car's skid marks to be 135 feet long. Approximately how fast was the car traveling?

The formula $v = \sqrt{2.7r}$ models the maximum safe speed, v, in miles per hour, at which a car can travel on a curved road with radius of curvature r, in feet.

19. A highway crew measures the radius of curvature at an exit ramp on a highway as 430 feet. What is the maximum safe speed?

20. A highway crew measures the radius of curvature at a tight corner on a highway as 900 feet. What is the maximum safe speed?

21. A drainage canal has a cross-section in the shape of a parabola. Suppose that the canal is 10 feet deep and 20 feet wide at the top. If the water depth in the ditch is 5 feet, how wide is the surface of the water in the ditch? [UW]

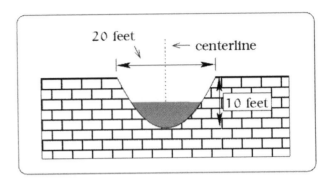

22. Brooke is located 5 miles out from the nearest point A along a straight shoreline in her sea kayak. Hunger strikes and she wants to make it to Kono's for lunch; see picture. Brooke can paddle 2 mph and walk 4 mph. [UW]

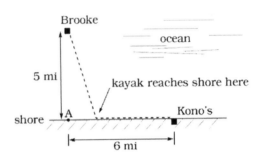

a. If she paddles along a straight line course to the shore, find an expression that computes the total time to reach lunch in terms of the location where Brooke beaches her kayak.

b. Determine the total time to reach Kono's if she paddles directly to the point A.

c. Determine the total time to reach Kono's if she paddles directly to Kono's.

d. Do you think your answer to b or c is the minimum time required for Brooke to reach lunch?

e. Determine the total time to reach Kono's if she paddles directly to a point on the shore half way between point A and Kono's. How does this time compare to the times in parts b or c? Do you need to modify your answer to part d?

23. Clovis is standing at the edge of a dropoff, which slopes 4 feet downward from him for every 1 horizontal foot. He launches a small model rocket from where he is standing. With the origin of the coordinate system located where he is standing, and the x-axis extending horizontally, the path of the rocket is described by the formula $y = -2x^2 + 120x$. [UW]

a. Give a function $h = f(x)$ relating the height h of the rocket above the sloping ground to its x-coordinate.

b. Find the maximum height of the rocket above the sloping ground. What is its x-coordinate when it is at its maximum height?

c. Clovis measures the height h of the rocket above the sloping ground while it is going up. Give a function $x = g(h)$ relating the x-coordinate of the rocket to h.

d. Does the function from (c) still work when the rocket is going down? Explain.

24. A trough has a semicircular cross section with a radius of 5 feet. Water starts flowing into the trough in such a way that the depth of the water is increasing at a rate of 2 inches per hour. [UW]

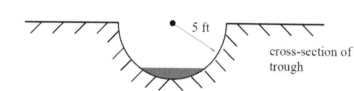

water

5 ft

cross-section of trough

a. Give a function $w = f(t)$ relating the width w of the surface of the water to the time t, in hours. Make sure to specify the domain and compute the range too.

b. After how many hours will the surface of the water have width of 6 feet?

c. Give a function $t = f^{-1}(w)$ relating the time to the width of the surface of the water. Make sure to specify the domain and compute the range too.

Chapter 4:
Exponential and Logarithmic Functions

Section 4.1 Exponential Functions

India is the second most populous country in the world, with a population in 2008 of about 1.14 billion people. The population is growing by about 1.34% each year[1]. We might ask if we can find a formula to model the population, P, as a function of time, t, in years after 2008, if the population continues to grow at this rate.

In linear growth, we had a constant rate of change – a constant *number* that the output increased for each increase in input. For example, in the equation $f(x) = 3x + 4$, the slope tells us the output increases by three each time the input increases by one. This population scenario is different – we have a *percent* rate of change rather than a constant number of people as our rate of change.

To see the significance of this difference consider these two companies:

Company *A* has 100 stores, and expands by opening 50 new stores a year

Company *B* has 100 stores, and expands by increasing the number of stores by 50% of their total each year.

Looking at a few years of growth for these companies:

---

[1] World Bank, World Development Indicators, as reported on http://www.google.com/publicdata, retrieved August 20, 2010

Year	Stores, company A		Stores, company B
0	100	Starting with 100 each	100
1	100 + 50 = 150	They both grow by 50 stores in the first year.	100 + 50% of 100 100 + 0.50(100) = 150
2	150 + 50 = 200	Store A grows by 50, Store B grows by 75	150 + 50% of 150 150 + 0.50(150) = 225
3	200 + 50 = 250	Store A grows by 50, Store B grows by 112.5	225 + 50% of 225 225 + 0.50(225) = 337.5

Notice that with the percent growth, each year the company is grows by 50% of the current year's total, so as the company grows larger, the number of stores added in a year grows as well.

To try to simplify the calculations, notice that after 1 year the number of stores for company B was:
$100 + 0.50(100)$ or equivalently by factoring
$100(1 + 0.50) = 150$

We can think of this as "the new number of stores is the original 100% plus another 50%".

After 2 years, the number of stores was:
$150 + 0.50(150)$ or equivalently by factoring
$150(1 + 0.50)$ now recall the 150 came from 100(1+0.50). Substituting that,
$100(1 + 0.50)(1 + 0.50) = 100(1 + 0.50)^2 = 225$

After 3 years, the number of stores was:
$225 + 0.50(225)$ or equivalently by factoring
$225(1 + 0.50)$ now recall the 225 came from $100(1 + 0.50)^2$. Substituting that,
$100(1 + 0.50)^2 (1 + 0.50) = 100(1 + 0.50)^3 = 337.5$

From this, we can generalize, noticing that to show a 50% increase, each year we multiply by a factor of (1+0.50), so after n years, our equation would be
$B(n) = 100(1 + 0.50)^n$

In this equation, the 100 represented the initial quantity, and the 0.50 was the percent growth rate. Generalizing further, we arrive at the general form of exponential functions.

Exponential Function

An **exponential growth or decay function** is a function that grows or shrinks at a constant percent growth rate. The equation can be written in the form

$$f(x) = a(1+r)^x \quad \text{or} \quad f(x) = ab^x \quad \text{where } b = 1+r$$

Where

a is the initial or starting value of the function

r is the percent growth or decay rate, written as a decimal

b is the growth factor or growth multiplier. Since powers of negative numbers behave strangely, we limit b to positive values.

To see more clearly the difference between exponential and linear growth, compare the two tables and graphs below, which illustrate the growth of company A and B described above over a longer time frame if the growth patterns were to continue.

years	Company A	Company B
2	200	225
4	300	506
6	400	1139
8	500	2563
10	600	5767

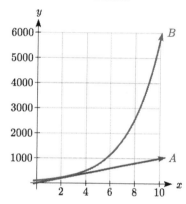

Example 1

Write an exponential function for India's population, and use it to predict the population in 2020.

At the beginning of the chapter we were given India's population of 1.14 billion in the year 2008 and a percent growth rate of 1.34%. Using 2008 as our starting time ($t = 0$), our initial population will be 1.14 billion. Since the percent growth rate was 1.34%, our value for r is 0.0134.

Using the basic formula for exponential growth $f(x) = a(1+r)^x$ we can write the formula, $f(t) = 1.14(1+0.0134)^t$

To estimate the population in 2020, we evaluate the function at $t = 12$, since 2020 is 12 years after 2008.

$f(12) = 1.14(1+0.0134)^{12} \approx 1.337$ billion people in 2020

Example 2

A certificate of deposit (CD) is a type of savings account offered by banks, typically offering a higher interest rate in return for a fixed length of time you will leave your money invested. If a bank offers a 24 month CD with an annual interest rate of 1.2% compounded monthly, how much will a $1000 investment grow to over those 24 months?

First, we must notice that the interest rate is an annual rate, but is compounded monthly, meaning interest is calculated and added to the account monthly. To find the monthly interest rate, we divide the annual rate of 1.2% by 12 since there are 12 months in a year: $1.2\%/12 = 0.1\%$. Each month we will earn 0.1% interest. From this, we can set up an exponential function, with our initial amount of $1000 and a growth rate of $r = 0.001$, and our input m measured in months.

$$f(m) = 1000\left(1 + \frac{.012}{12}\right)^m$$

$$f(m) = 1000(1 + 0.001)^m$$

After 24 months, the account will have grown to $f(24) = 1000(1 + 0.001)^{24} = \$1024.28$

In all the preceding examples, we saw exponential growth. Exponential functions can also be used to model quantities that are decreasing at a constant percent rate. An example of this is radioactive decay, a process in which radioactive isotopes of certain atoms transform to an atom of a different type, causing a percentage decrease of the original material over time.

Example 3

Bismuth-210 is an isotope that radioactively decays by about 13% each day, meaning 13% of the remaining Bismuth-210 transforms into another atom (polonium-210 in this case) each day. If you begin with 100 mg of Bismuth-210, how much remains after one week?

With radioactive decay, instead of the quantity increasing at a percent rate, the quantity is decreasing at a percent rate. Our initial quantity is $a = 100$ mg, and our growth rate will be negative 13%, since we are decreasing: $r = -0.13$. This gives the equation:
$Q(d) = 100(1 - 0.13)^d = 100(0.87)^d$
This can also be explained by recognizing that if 13% decays, then 87 % remains.

After one week, 7 days, the quantity remaining would be
$Q(7) = 100(0.87)^7 = 37.73$ mg of Bismuth-210 remains.

Try it Now
3. A population of 1000 is decreasing 3% each year. Find the population in 30 years.

Example 4

$T(q)$ represents the total number of Android smart phone contracts, in thousands, held by a certain Verizon store region measured quarterly since January 1, 2016, Interpret all the parts of the equation $T(2) = 86(1.64)^2 = 231.3056$.

Interpreting this from the basic exponential form, we know that 86 is our initial value. This means that on Jan. 1, 2016 this region had 86,000 Android smart phone contracts. Since $b = 1 + r = 1.64$, we know that every quarter the number of smart phone contracts grows by 64%. $T(2) = 231.3056$ means that in the 2nd quarter (or at the end of the second quarter) there were approximately 231,306 Android smart phone contracts.

Finding Equations of Exponential Functions

In the previous examples, we were able to write equations for exponential functions since we knew the initial quantity and the growth rate. If we do not know the growth rate, but instead know only some input and output pairs of values, we can still construct an exponential function.

Example 5

In 2009, 80 deer were reintroduced into a wildlife refuge area from which the population had previously been hunted to elimination. By 2015, the population had grown to 180 deer. If this population grows exponentially, find a formula for the function.

By defining our input variable to be t, years after 2009, the information listed can be written as two input-output pairs: (0,80) and (6,180). Notice that by choosing our input variable to be measured as years after the first year value provided, we have effectively "given" ourselves the initial value for the function: $a = 80$. This gives us an equation of the form

$$f(t) = 80b^t.$$

Substituting in our second input-output pair allows us to solve for b:

$180 = 80b^6$ Divide by 80

$b^6 = \dfrac{180}{80} = \dfrac{9}{4}$ Take the 6$^\text{th}$ root of both sides.

$b = \sqrt[6]{\dfrac{9}{4}} = 1.1447$

This gives us our equation for the population:

$$f(t) = 80(1.1447)^t$$

Recall that since $b = 1+r$, we can interpret this to mean that the population growth rate is $r = 0.1447$, and so the population is growing by about 14.47% each year.

In this example, you could also have used (9/4)^(1/6) to evaluate the 6$^\text{th}$ root if your calculator doesn't have an nth root button.

In the previous example, we chose to use the $f(x) = ab^x$ form of the exponential function rather than the $f(x) = a(1+r)^x$ form. This choice was entirely arbitrary – either form would be fine to use.

When finding equations, the value for b or r will usually have to be rounded to be written easily. To preserve accuracy, it is important to not over-round these values. Typically, you want to be sure to preserve at least 3 significant digits in the growth rate. For example, if your value for b was 1.00317643, you would want to round this no further than to 1.00318.

In the previous example, we were able to "give" ourselves the initial value by clever definition of our input variable. Next, we consider a situation where we can't do this.

Example 6

Find a formula for an exponential function passing through the points (-2,6) and (2,1).

Since we don't have the initial value, we will take a general approach that will work for any function form with unknown parameters: we will substitute in both given input-output pairs in the function form $f(x) = ab^x$ and solve for the unknown values, a and b.

Substituting in (-2, 6) gives $6 = ab^{-2}$

Substituting in (2, 1) gives $1 = ab^2$

We now solve these as a system of equations. To do so, we could try a substitution approach, solving one equation for a variable, then substituting that expression into the second equation.

Solving $6 = ab^{-2}$ for a:

$$a = \frac{6}{b^{-2}} = 6b^2$$

In the second equation, $1 = ab^2$, we substitute the expression above for a:

$1 = (6b^2)b^2$

$1 = 6b^4$

$$\frac{1}{6} = b^4$$

$$b = \sqrt[4]{\frac{1}{6}} \approx 0.6389$$

Going back to the equation $a = 6b^2$ lets us find a:

$a = 6b^2 = 6(0.6389)^2 = 2.4492$

Putting this together gives the equation $f(x) = 2.4492(0.6389)^x$

Try it Now

4. Given the two points (1, 3) and (2, 4.5) find the equation of an exponential function that passes through these two points.

Example 7

Find an equation for the exponential function graphed.

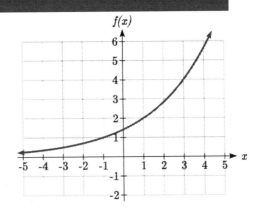

The initial value for the function is not clear in this graph, so we will instead work using two clearer points. There are three clear points: (-1, 1), (1, 2), and (3, 4). As we saw in the last example, two points are sufficient to find the equation for a standard exponential, so we will use the latter two points.

Substituting in (1,2) gives $2 = ab^1$

Substituting in (3,4) gives $4 = ab^3$

Solving the first equation for a gives $a = \dfrac{2}{b}$.

Substituting this expression for a into the second equation:

$4 = ab^3$

$4 = \dfrac{2}{b}b^3 = \dfrac{2b^3}{b}$ Simplify the right-hand side

$4 = 2b^2$

$2 = b^2$

$b = \pm\sqrt{2}$

Since we restrict ourselves to positive values of b, we will use $b = \sqrt{2}$. We can then go back and find a:

$a = \dfrac{2}{b} = \dfrac{2}{\sqrt{2}} = \sqrt{2}$

This gives us a final equation of $f(x) = \sqrt{2}(\sqrt{2})^x$.

Compound Interest

In the bank certificate of deposit (CD) example earlier in the section, we encountered compound interest. Typically bank accounts and other savings instruments in which earnings are reinvested, such as mutual funds and retirement accounts, utilize compound interest. The term *compounding* comes from the behavior that interest is earned not on the original value, but on the accumulated value of the account.

In the example from earlier, the interest was compounded monthly, so we took the annual interest rate, usually called the **nominal rate** or **annual percentage rate (APR)** and divided by 12, the number of compounds in a year, to find the monthly interest. The exponent was then measured in months.

Generalizing this, we can form a general formula for compound interest. If the APR is written in decimal form as r, and there are k compounding periods per year, then the interest per compounding period will be r/k. Likewise, if we are interested in the value after t years, then there will be kt compounding periods in that time.

Compound Interest Formula

Compound Interest can be calculated using the formula

$$A(t) = a\left(1 + \frac{r}{k}\right)^{kt}$$

Where

$A(t)$ is the account value

t is measured in years

a is the starting amount of the account, often called the principal

r is the annual percentage rate (APR), also called the nominal rate

k is the number of compounding periods in one year

Example 8

If you invest $3,000 in an investment account paying 3% interest compounded quarterly, how much will the account be worth in 10 years?

Since we are starting with $3000, $a = 3000$
Our interest rate is 3%, so $r = 0.03$
Since we are compounding quarterly, we are compounding 4 times per year, so $k = 4$
We want to know the value of the account in 10 years, so we are looking for $A(10)$, the value when $t = 10$.

$$A(10) = 3000\left(1 + \frac{0.03}{4}\right)^{4(10)} = \$4045.05$$

The account will be worth $4045.05 in 10 years.

Example 9

A 529 plan is a college savings plan in which a relative can invest money to pay for a child's later college tuition, and the account grows tax free. If Lily wants to set up a 529 account for her new granddaughter, wants the account to grow to $40,000 over 18 years, and she believes the account will earn 6% compounded semi-annually (twice a year), how much will Lily need to invest in the account now?

Since the account is earning 6%, $r = 0.06$
Since interest is compounded twice a year, $k = 2$

In this problem, we don't know how much we are starting with, so we will be solving for a, the initial amount needed. We do know we want the end amount to be $40,000, so we will be looking for the value of a so that $A(18) = 40,000$.

$$40,000 = A(18) = a\left(1 + \frac{0.06}{2}\right)^{2(18)}$$

$$40,000 = a(2.8983)$$

$$a = \frac{40,000}{2.8983} \approx \$13,801$$

Lily will need to invest $13,801 to have $40,000 in 18 years.

Try it now
5. Recalculate example 2 from above with quarterly compounding.

Because of compounding throughout the year, with compound interest the actual increase in a year is *more* than the annual percentage rate. If $1,000 were invested at 10%, the table below shows the value after 1 year at different compounding frequencies:

Frequency	Value after 1 year
Annually	$1100
Semiannually	$1102.50
Quarterly	$1103.81
Monthly	$1104.71
Daily	$1105.16

If we were to compute the actual percentage increase for the daily compounding, there was an increase of $105.16 from an original amount of $1,000, for a percentage increase of $\frac{105.16}{1000} = 0.10516 = 10.516\%$ increase. This quantity is called the **annual percentage yield (APY)**.

Notice that given any starting amount, the amount after 1 year would be

$$A(1) = a\left(1 + \frac{r}{k}\right)^k.$$ To find the total change, we would subtract the original amount, then

to find the percentage change we would divide that by the original amount:

$$\frac{a\left(1 + \frac{r}{k}\right)^k - a}{a} = \left(1 + \frac{r}{k}\right)^k - 1$$

Annual Percentage Yield

The **annual percentage yield** is the actual percent a quantity increases in one year. It can be calculated as

$$APY = \left(1 + \frac{r}{k}\right)^k - 1$$

This is equivalent to finding the value of $1 after 1 year, and subtracting the original dollar.

Example 10

Bank A offers an account paying 1.2% compounded quarterly. Bank B offers an account paying 1.1% compounded monthly. Which is offering a better rate?

We can compare these rates using the annual percentage yield – the actual percent increase in a year.

Bank A: $APY = \left(1 + \dfrac{0.012}{4}\right)^4 - 1 = 0.012054 = 1.2054\%$

Bank B: $APY = \left(1 + \dfrac{0.011}{12}\right)^{12} - 1 = 0.011056 = 1.1056\%$

Bank B's monthly compounding is not enough to catch up with Bank A's better APR. Bank A offers a better rate.

A Limit to Compounding

As we saw earlier, the amount we earn increases as we increase the compounding frequency. The table, though, shows that the increase from annual to semi-annual compounding is larger than the increase from monthly to daily compounding. This might lead us to believe that although increasing the frequency of compounding will increase our result, there is an upper limit to this process.

To see this, let us examine the value of $1 invested at 100% interest for 1 year.

Frequency	Value
Annual	$2
Quarterly	$2.441406
Monthly	$2.613035
Daily	$2.714567
Hourly	$2.718127
Once per minute	$2.718279
Once per second	$2.718282

These values do indeed appear to be approaching an upper limit. This value ends up being so important that it gets represented by its own letter, much like how π represents a number.

Euler's Number: e

e is the letter used to represent the value that $\left(1+\dfrac{1}{k}\right)^k$ approaches as k gets big.

$e \approx 2.718282$

Because e is often used as the base of an exponential, most scientific and graphing calculators have a button that can calculate powers of e, usually labeled e^x. Some computer software instead defines a function $exp(x)$, where $exp(x) = e^x$.

Because e arises when the time between compounds becomes very small, e allows us to define **continuous growth** and allows us to define a new toolkit function, $f(x) = e^x$.

Continuous Growth Formula

Continuous Growth can be calculated using the formula

$f(x) = ae^{rx}$

where

a is the starting amount

r is the continuous growth rate

This type of equation is commonly used when describing quantities that change more or less continuously, like chemical reactions, growth of large populations, and radioactive decay.

Example 11

Radon-222 decays at a continuous rate of 17.3% per day. How much will 100mg of Radon-222 decay to in 3 days?

Since we are given a continuous decay rate, we use the continuous growth formula. Since the substance is decaying, we know the growth rate will be negative: $r = -0.173$
$f(3) = 100\,e^{-0.173(3)} \approx 59.512$ mg of Radon-222 will remain.

Try it Now

6. Interpret the following: $S(t) = 20e^{0.12t}$ if $S(t)$ represents the growth of a substance in grams, and time is measured in days.

Continuous growth is also often applied to compound interest, allowing us to talk about continuous compounding.

Example 12

If $1000 is invested in an account earning 10% compounded continuously, find the value after 1 year.

Here, the continuous growth rate is 10%, so $r = 0.10$. We start with $1000, so $a = 1000$. To find the value after 1 year,
$f(1) = 1000\,e^{0.10(1)} \approx \$1105.17$

Notice this is a $105.17 increase for the year. As a percent increase, this is
$\dfrac{105.17}{1000} = 0.10517 = 10.517\%$ increase over the original $1000.

Notice that this value is slightly larger than the amount generated by daily compounding in the table computed earlier.

The continuous growth rate is like the nominal growth rate (or APR) – it reflects the growth rate before compounding takes effect. This is different than the annual growth rate used in the formula $f(x) = a(1 + r)^x$, which is like the annual percentage yield – it reflects the *actual* amount the output grows in a year.

While the continuous growth rate in the example above was 10%, the actual annual yield was 10.517%. This means we could write two different looking but equivalent formulas for this account's growth:
$f(t) = 1000e^{0.10t}$ using the 10% continuous growth rate
$f(t) = 1000(1.10517)^t$ using the 10.517% actual annual yield rate.

Important Topics of this Section
Percent growth
Exponential functions
Finding formulas
Interpreting equations
Graphs
Exponential Growth & Decay
Compound interest
Annual Percent Yield
Continuous Growth

Try it Now Answers

1. A & C are exponential functions, they grow by a % not a constant number.

2. B(t) is growing faster ($r = 0.075 > 0.05$), but after 3 years A(t) still has a higher account balance

3. $P(t) = 1000(1 - 0.03)^t = 1000(0.97)^t$
   $P(30) = 1000(0.97)^{30} = 401.0071$

4. $3 = ab^1$, so $a = \dfrac{3}{b}$,

   $4.5 = ab^2$, so $4.5 = \dfrac{3}{b}b^2$. $4.5 = 3b$

   $b = 1.5$. $a = \dfrac{3}{1.5} = 2$

   $f(x) = 2(1.5)^x$

5. 24 months = 2 years. $1000\left(1 + \dfrac{.012}{4}\right)^{4(2)} = \$1024.25$

6. An initial substance weighing 20g is growing at a continuous rate of 12% per day.

Section 4.1 Exercises

For each table below, could the table represent a function that is linear, exponential, or neither?

1.

x	1	2	3	4
$f(x)$	70	40	10	-20

2.

x	1	2	3	4
$g(x)$	40	32	26	22

3.

x	1	2	3	4
$h(x)$	70	49	34.3	24.01

4.

x	1	2	3	4
$k(x)$	90	80	70	60

5.

x	1	2	3	4
$m(x)$	80	61	42.9	25.61

6.

x	1	2	3	4
$n(x)$	90	81	72.9	65.61

7. A population numbers 11,000 organisms initially and grows by 8.5% each year. Write an exponential model for the population.

8. A population is currently 6,000 and has been increasing by 1.2% each day. Write an exponential model for the population.

9. The fox population in a certain region has an annual growth rate of 9 percent per year. It is estimated that the population in the year 2010 was 23,900. Estimate the fox population in the year 2018.

10. The amount of area covered by blackberry bushes in a park has been growing by 12% each year. It is estimated that the area covered in 2009 was 4,500 square feet. Estimate the area that will be covered in 2020.

11. A vehicle purchased for $32,500 depreciates at a constant rate of 5% each year. Determine the approximate value of the vehicle 12 years after purchase.

12. A business purchases $125,000 of office furniture which depreciates at a constant rate of 12% each year. Find the residual value of the furniture 6 years after purchase.

Find a formula for an exponential function passing through the two points.

13. $(0,6), (3,750)$ 　　　　　　　　14. $(0,3), (2,75)$

15. $(0,2000), (2,20)$ 　　　　　　16. $(0,9000), (3,72)$

17. $\left(-1,\dfrac{3}{2}\right), (3,24)$ 　　　　　18. $\left(-1,\dfrac{2}{5}\right), (1,10)$

19. $(-2,6), (3,1)$ 　　　　　　　　20. $(-3,4), (3,2)$

21. $(3,1), (5,4)$ 　　　　　　　　22. $(2,5), (6,9)$

23. A radioactive substance decays exponentially. A scientist begins with 100 milligrams of a radioactive substance. After 35 hours, 50 mg of the substance remains. How many milligrams will remain after 54 hours?

24. A radioactive substance decays exponentially. A scientist begins with 110 milligrams of a radioactive substance. After 31 hours, 55 mg of the substance remains. How many milligrams will remain after 42 hours?

25. A house was valued at $110,000 in the year 1985. The value appreciated to $145,000 by the year 2005. What was the annual growth rate between 1985 and 2005? Assume that the house value continues to grow by the same percentage. What did the value equal in the year 2010?

26. An investment was valued at $11,000 in the year 1995. The value appreciated to $14,000 by the year 2008. What was the annual growth rate between 1995 and 2008? Assume that the value continues to grow by the same percentage. What did the value equal in the year 2012?

27. A car was valued at $38,000 in the year 2003. The value depreciated to $11,000 by the year 2009. Assume that the car value continues to drop by the same percentage. What was the value in the year 2013?

28. A car was valued at $24,000 in the year 2006. The value depreciated to $20,000 by the year 2009. Assume that the car value continues to drop by the same percentage. What was the value in the year 2014?

29. If $4,000 is invested in a bank account at an interest rate of 7 per cent per year, find the amount in the bank after 9 years if interest is compounded annually, quarterly, monthly, and continuously.

30. If $6,000 is invested in a bank account at an interest rate of 9 per cent per year, find the amount in the bank after 5 years if interest is compounded annually, quarterly, monthly, and continuously.

31. Find the annual percentage yield (APY) for a savings account with annual percentage rate of 3% compounded quarterly.

32. Find the annual percentage yield (APY) for a savings account with annual percentage rate of 5% compounded monthly.

33. A population of bacteria is growing according to the equation $P(t)=1600e^{0.21t}$, with t measured in years. Estimate when the population will exceed 7569.

34. A population of bacteria is growing according to the equation $P(t)=1200e^{0.17t}$, with t measured in years. Estimate when the population will exceed 3443.

35. In 1968, the U.S. minimum wage was $1.60 per hour. In 1976, the minimum wage was $2.30 per hour. Assume the minimum wage grows according to an exponential model $w(t)$, where t represents the time in years after 1960. [UW]
    a. Find a formula for $w(t)$.
    b. What does the model predict for the minimum wage in 1960?
    c. If the minimum wage was $5.15 in 1996, is this above, below or equal to what the model predicts?

36. In 1989, research scientists published a model for predicting the cumulative number of AIDS cases (in thousands) reported in the United States: $a(t)=155\left(\dfrac{t-1980}{10}\right)^3$, where t is the year. This paper was considered a "relief", since there was a fear the correct model would be of exponential type. Pick two data points predicted by the research model $a(t)$ to construct a new exponential model $b(t)$ for the number of cumulative AIDS cases. Discuss how the two models differ and explain the use of the word "relief." [UW]

37. You have a chess board as pictured, with squares numbered 1 through 64. You also have a huge change jar with an unlimited number of dimes. On the first square you place one dime. On the second square you stack 2 dimes. Then you continue, always doubling the number from the previous square. [UW]

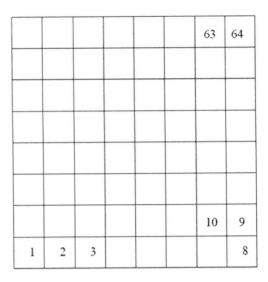

   a. How many dimes will you have stacked on the 10th square?

   b. How many dimes will you have stacked on the nth square?

   c. How many dimes will you have stacked on the 64th square?

   d. Assuming a dime is 1 mm thick, how high will this last pile be?

   e. The distance from the earth to the sun is approximately 150 million km. Relate the height of the last pile of dimes to this distance.

Section 4.2 Graphs of Exponential Functions

Like with linear functions, the graph of an exponential function is determined by the values for the parameters in the function's formula.

To get a sense for the behavior of exponentials, let us begin by looking more closely at the function $f(x) = 2^x$. Listing a table of values for this function:

x	-3	-2	-1	0	1	2	3
$f(x)$	$\dfrac{1}{8}$	$\dfrac{1}{4}$	$\dfrac{1}{2}$	1	2	4	8

Notice that:
1) This function is positive for all values of x.
2)   As x increases, the function grows faster and faster (the rate of change increases).
3) As x decreases, the function values grow smaller, approaching zero.
4) This is an example of exponential growth.

Looking at the function $g(x) = \left(\dfrac{1}{2}\right)^x$

x	-3	-2	-1	0	1	2	3
$g(x)$	8	4	2	1	$\dfrac{1}{2}$	$\dfrac{1}{4}$	$\dfrac{1}{8}$

Note this function is also positive for all values of x, but in this case grows as x decreases, and decreases towards zero as x increases. This is an example of exponential decay. You may notice from the table that this function appears to be the horizontal reflection of the $f(x) = 2^x$ table. This is in fact the case:

$$f(-x) = 2^{-x} = (2^{-1})^x = \left(\frac{1}{2}\right)^x = g(x)$$

Looking at the graphs also confirms this relationship.

Consider a function of the form $f(x) = ab^x$.
Since a, which we called the initial value in the last section, is the function value at an input of zero, a will give us the vertical intercept of the graph.

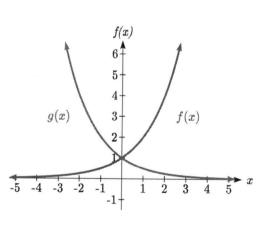

From the graphs above, we can see that an exponential graph will have a horizontal asymptote on one side of the graph, and can either increase or decrease, depending upon the growth factor. This horizontal asymptote will also help us determine the long run behavior and is easy to determine from the graph.

The graph will grow when the growth rate is positive, which will make the growth factor b larger than one. When it's negative, the growth factor will be less than one.

Graphical Features of Exponential Functions

Graphically, in the function $f(x) = ab^x$

a is the vertical intercept of the graph

b determines the rate at which the graph grows. When a is positive,

the function will increase if $b > 1$

the function will decrease if $0 < b < 1$

The graph will have a horizontal asymptote at $y = 0$

The graph will be concave up if $a > 0$; concave down if $a < 0$.

The domain of the function is all real numbers

The range of the function is $(0, \infty)$

When sketching the graph of an exponential function, it can be helpful to remember that the graph will pass through the points $(0, a)$ and $(1, ab)$.

The value b will determine the function's long run behavior:

If $b > 1$, as $x \to \infty$, $f(x) \to \infty$ and as $x \to -\infty$, $f(x) \to 0$.

If $0 < b < 1$, as $x \to \infty$, $f(x) \to 0$ and as $x \to -\infty$, $f(x) \to \infty$.

Example 1

Sketch a graph of $f(x) = 4\left(\dfrac{1}{3}\right)^x$

This graph will have a vertical intercept at $(0,4)$, and pass through the point $\left(1, \dfrac{4}{3}\right)$. Since $b < 1$, the graph will be decreasing towards zero. Since $a > 0$, the graph will be concave up.

We can also see from the graph the long run behavior: as $x \to \infty$, $f(x) \to 0$ and as $x \to -\infty$, $f(x) \to \infty$.

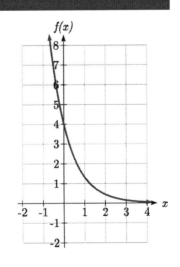

To get a better feeling for the effect of a and b on the graph, examine the sets of graphs below. The first set shows various graphs, where a remains the same and we only change the value for b.

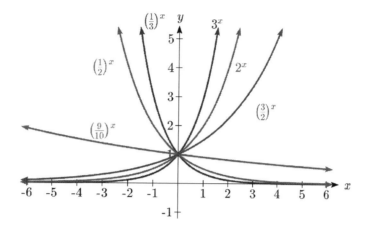

Notice that the closer the value of b is to 1, the less steep the graph will be.

In the next set of graphs, a is altered and our value for b remains the same.

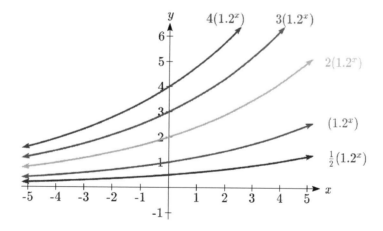

Notice that changing the value for a changes the vertical intercept. Since a is multiplying the b^x term, a acts as a vertical stretch factor, not as a shift. Notice also that the long run behavior for all of these functions is the same because the growth factor did not change and none of these a values introduced a vertical flip.

Example 2

Match each equation with its graph.

$f(x) = 2(1.3)^x$

$g(x) = 2(1.8)^x$

$h(x) = 4(1.3)^x$

$k(x) = 4(0.7)^x$

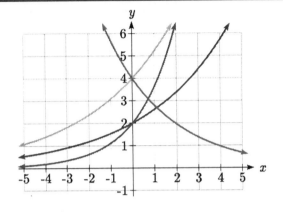

The graph of $k(x)$ is the easiest to identify, since it is the only equation with a growth factor less than one, which will produce a decreasing graph. The graph of $h(x)$ can be identified as the only growing exponential function with a vertical intercept at (0,4). The graphs of $f(x)$ and $g(x)$ both have a vertical intercept at (0,2), but since $g(x)$ has a larger growth factor, we can identify it as the graph increasing faster.

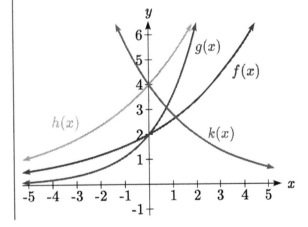

Try it Now

1. Graph the following functions on the same axis:

$f(x) = (2)^x$; $g(x) = 2(2)^x$; $h(x) = 2(1/2)^x$.

Transformations of Exponential Graphs

While exponential functions can be transformed following the same rules as any function, there are a few interesting features of transformations that can be identified. The first was seen at the beginning of the section – that a horizontal reflection is equivalent to a change in the growth factor. Likewise, since a is itself a stretch factor, a vertical stretch of an exponential corresponds with a change in the initial value of the function.

Next consider the effect of a horizontal shift on an exponential function. Shifting the function $f(x) = 3(2)^x$ four units to the left would give $f(x+4) = 3(2)^{x+4}$. Employing exponent rules, we could rewrite this:

$$f(x+4) = 3(2)^{x+4} = 3(2)^x (2^4) = 48(2)^x$$

Interestingly, it turns out that a horizontal shift of an exponential function corresponds with a change in initial value of the function.

Lastly, consider the effect of a vertical shift on an exponential function. Shifting $f(x) = 3(2)^x$ down 4 units would give the equation $f(x) = 3(2)^x - 4$.

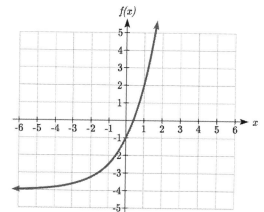

Graphing that, notice it is substantially different than the basic exponential graph. Unlike a basic exponential, this graph does not have a horizontal asymptote at $y = 0$; due to the vertical shift, the horizontal asymptote has also shifted to $y = -4$. We can see that as $x \to \infty$, $f(x) \to \infty$ and as $x \to -\infty$, $f(x) \to -4$.

We have determined that a vertical shift is the only transformation of an exponential function that changes the graph in a way that cannot be achieved by altering the parameters a and b in the basic exponential function $f(x) = ab^x$.

Transformations of Exponentials
Any transformed exponential can be written in the form $$f(x) = ab^x + c$$ where $y = c$ is the horizontal asymptote.

Note that, due to the shift, the vertical intercept is shifted to $(0, a+c)$.

Try it Now

2. Write the equation and graph the exponential function described as follows:
$f(x) = e^x$ is vertically stretched by a factor of 2, flipped across the y axis and shifted up 4 units.

Example 3

Sketch a graph of $f(x) = -3\left(\dfrac{1}{2}\right)^x + 4$.

Notice that in this exponential function, the negative in the stretch factor -3 will cause a vertical reflection, and the vertical shift up 4 will move the horizontal asymptote to $y = 4$. Sketching this as a transformation of $g(x) = \left(\dfrac{1}{2}\right)^x$,

The basic $g(x) = \left(\dfrac{1}{2}\right)^x$ Vertically reflected and stretched by 3

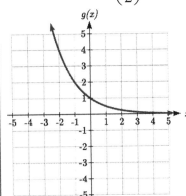
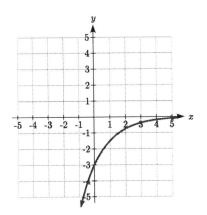

Vertically shifted up four units

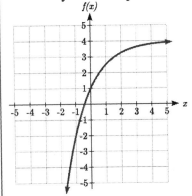

Notice that while the domain of this function is unchanged, due to the reflection and shift, the range of this function is $(-\infty, 4)$.

As $x \to \infty$, $f(x) \to 4$ and as $x \to -\infty$, $f(x) \to -\infty$.

Functions leading to graphs like the one above are common as models for learning and models of growth approaching a limit.

Example 4

Find an equation for the function graphed.

Looking at this graph, it appears to have a horizontal asymptote at $y = 5$, suggesting an equation of the form $f(x) = ab^x + 5$. To find values for a and b, we can identify two other points on the graph. It appears the graph passes through $(0,2)$ and $(-1,3)$, so we can use those points. Substituting in $(0,2)$ allows us to solve for a.

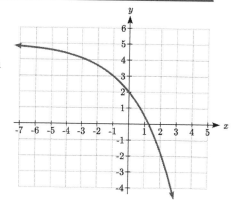

$2 = ab^0 + 5$

$2 = a + 5$

$a = -3$

Substituting in $(-1,3)$ allows us to solve for b

$3 = -3b^{-1} + 5$

$-2 = \dfrac{-3}{b}$

$-2b = -3$

$b = \dfrac{3}{2} = 1.5$

The final formula for our function is $f(x) = -3(1.5)^x + 5$.

Try it Now

3. Given the graph of the transformed exponential function, find a formula and describe the long run behavior.

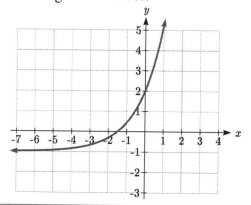

Important Topics of this Section
Graphs of exponential functions
Intercept
Growth factor
Exponential Growth
Exponential Decay
Horizontal intercepts
Long run behavior
Transformations

Try it Now Answers

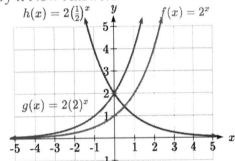

1.

2. $f(x) = -2e^x + 4$

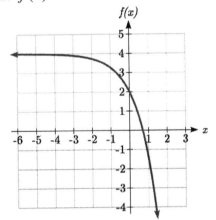

3. Horizontal asymptote at $y = -1$, so $f(x) = ab^x - 1$. Substitute (0, 2) to find $a = 3$.

Substitute (1,5) to find $5 = 3b^1 - 1$, $b = 2$.

$f(x) = 3(2^x) - 1$ or $f(x) = 3(.5)^{-x} - 1$

As $x \to \infty$, $f(x) \to \infty$ and as $x \to -\infty$, $f(x) \to -1$

Section 4.2 Exercises

Match each function with one of the graphs below.

1. $f(x) = 2(0.69)^x$

2. $f(x) = 2(1.28)^x$

3. $f(x) = 2(0.81)^x$

4. $f(x) = 4(1.28)^x$

5. $f(x) = 2(1.59)^x$

6. $f(x) = 4(0.69)^x$

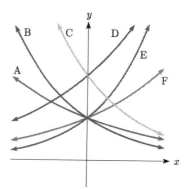

If all the graphs to the right have equations with form
$f(x) = ab^x$,

7. Which graph has the largest value for b?

8. Which graph has the smallest value for b?

9. Which graph has the largest value for a?

10. Which graph has the smallest value for a?

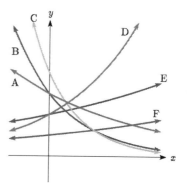

Sketch a graph of each of the following transformations of $f(x) = 2^x$

11. $f(x) = 2^{-x}$ 12. $g(x) = -2^x$

13. $h(x) = 2^x + 3$ 14. $f(x) = 2^x - 4$

15. $f(x) = 2^{x-2}$ 16. $k(x) = 2^{x-3}$

Starting with the graph of $f(x) = 4^x$, find a formula for the function that results from

17. Shifting $f(x)$ 4 units upwards

18. Shifting $f(x)$ 3 units downwards

19. Shifting $f(x)$ 2 units left

20. Shifting $f(x)$ 5 units right

21. Reflecting $f(x)$ about the x-axis

22. Reflecting $f(x)$ about the y-axis

Describe the long run behavior, as $x \to \infty$ and $x \to -\infty$ of each function

23. $f(x) = -5(4^x) - 1$

24. $f(x) = -2(3^x) + 2$

25. $f(x) = 3\left(\dfrac{1}{2}\right)^x - 2$

26. $f(x) = 4\left(\dfrac{1}{4}\right)^x + 1$

27. $f(x) = 3(4)^{-x} + 2$

28. $f(x) = -2(3)^{-x} - 1$

Find a formula for each function graphed as a transformation of $f(x) = 2^x$.

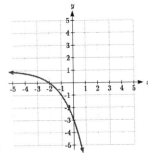

29.

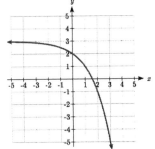

30.

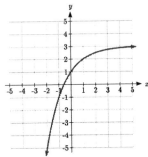

31.

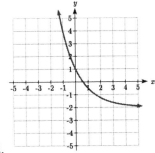

32.

Find an equation for the exponential function graphed.

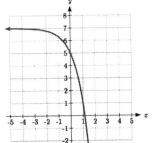

33.

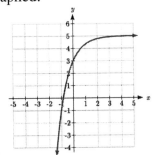

34.

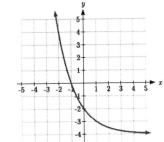

35.

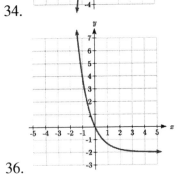

36.

Section 4.3 Logarithmic Functions

A population of 50 flies is expected to double every week, leading to a function of the form $f(x) = 50(2)^x$, where x represents the number of weeks that have passed. When will this population reach 500? Trying to solve this problem leads to:

$500 = 50(2)^x$ Dividing both sides by 50 to isolate the exponential

$10 = 2^x$

While we have set up exponential models and used them to make predictions, you may have noticed that solving exponential equations has not yet been mentioned. The reason is simple: none of the algebraic tools discussed so far are sufficient to solve exponential equations. Consider the equation $2^x = 10$ above. We know that $2^3 = 8$ and $2^4 = 16$, so it is clear that x must be some value between 3 and 4 since $g(x) = 2^x$ is increasing. We could use technology to create a table of values or graph to better estimate the solution.

From the graph, we could better estimate the solution to be around 3.3. This result is still fairly unsatisfactory, and since the exponential function is one-to-one, it would be great to have an inverse function. None of the functions we have already discussed would serve as an inverse function and so we must introduce a new function, named **log** as the inverse of an exponential function. Since exponential functions have different bases, we will define corresponding logarithms of different bases as well.

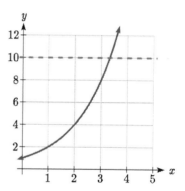

Logarithm

The logarithm (base b) function, written $\log_b(x)$, is the inverse of the exponential function (base b), b^x.

Since the logarithm and exponential are inverses, it follows that:

Properties of Logs: Inverse Properties

$\log_b(b^x) = x$

$b^{\log_b x} = x$

Recall from the definition of an inverse function that if $f(a) = c$, then $f^{-1}(c) = a$. Applying this to the exponential and logarithmic functions, we can convert between a logarithmic equation and its equivalent exponential.

Logarithm Equivalent to an Exponential

The statement $b^a = c$ is equivalent to the statement $\log_b(c) = a$.

Alternatively, we could show this by starting with the exponential function $c = b^a$, then taking the log base b of both sides, giving $\log_b(c) = \log_b b^a$. Using the inverse property of logs, we see that $\log_b(c) = a$.

Since log is a function, it is most correctly written as $\log_b(c)$, using parentheses to denote function evaluation, just as we would with $f(c)$. However, when the input is a single variable or number, it is common to see the parentheses dropped and the expression written as $\log_b c$.

Example 1

Write these exponential equations as logarithmic equations:

a) $2^3 = 8$ b) $5^2 = 25$ c) $10^{-4} = \dfrac{1}{10000}$

a) $2^3 = 8$ is equivalent to $\log_2(8) = 3$

b) $5^2 = 25$ is equivalent to $\log_5(25) = 2$

c) $10^{-4} = \dfrac{1}{10000}$ is equivalent to $\log_{10}\left(\dfrac{1}{10000}\right) = -4$

Example 2

Write these logarithmic equations as exponential equations:

a) $\log_6\left(\sqrt{6}\right) = \dfrac{1}{2}$ b) $\log_3(9) = 2$

a) $\log_6\left(\sqrt{6}\right) = \dfrac{1}{2}$ is equivalent to $6^{1/2} = \sqrt{6}$

b) $\log_3(9) = 2$ is equivalent to $3^2 = 9$

Try it Now

1. Write the exponential equation $4^2 = 16$ as a logarithmic equation.

By establishing the relationship between exponential and logarithmic functions, we can now solve basic logarithmic and exponential equations by rewriting.

Example 3

Solve $\log_4(x) = 2$ for x.

By rewriting this expression as an exponential, $4^2 = x$, so $x = 16$.

Example 4

Solve $2^x = 10$ for x.

By rewriting this expression as a logarithm, we get $x = \log_2(10)$.

While this does define a solution, and an exact solution at that, you may find it somewhat unsatisfying since it is difficult to compare this expression to the decimal estimate we made earlier. Also, giving an exact expression for a solution is not always useful – often we really need a decimal approximation to the solution. Luckily, this is a task calculators and computers are quite adept at. Unluckily for us, most calculators and computers will only evaluate logarithms of two bases. Happily, this ends up not being a problem, as we'll see briefly.

Common and Natural Logarithms

The **common log** is the logarithm with base 10, and is typically written $\log(x)$.

The **natural log** is the logarithm with base e, and is typically written $\ln(x)$.

Example 5

Evaluate $\log(1000)$ using the definition of the common log.

To evaluate $\log(1000)$, we can let $x = \log(1000)$, then rewrite into exponential form using the common log base of 10: $10^x = 1000$.

From this, we might recognize that 1000 is the cube of 10, so $x = 3$.

Values of the common log

number	number as exponential	log(number)
1000	10^3	3
100	10^2	2
10	10^1	1
1	10^0	0
0.1	10^{-1}	-1
0.01	10^{-2}	-2
0.001	10^{-3}	-3

We also can use the inverse property of logs to write $\log_{10}\left(10^3\right) = 3$.

Try it Now
2. Evaluate $\log(1000000)$.

Example 6

Evaluate $\ln\left(\sqrt{e}\right)$.

We can rewrite $\ln\left(\sqrt{e}\right)$ as $\ln\left(e^{1/2}\right)$. Since ln is a log base e, we can use the inverse property for logs: $\ln\left(e^{1/2}\right) = \log_e\left(e^{1/2}\right) = \dfrac{1}{2}$.

Example 7

Evaluate $\log(500)$ using your calculator or computer.

Using a computer, we can evaluate $\log(500) \approx 2.69897$

To utilize the common or natural logarithm functions to evaluate expressions like $\log_2(10)$, we need to establish some additional properties.

Properties of Logs: Exponent Property

$$\log_b\left(A^r\right) = r\log_b(A)$$

To show why this is true, we offer a proof:

Since the logarithmic and exponential functions are inverses, $b^{\log_b A} = A$.

Raising both sides to the r power, we get $A^r = \left(b^{\log_b A}\right)^r$.

Utilizing the exponential rule that states $\left(x^p\right)^q = x^{pq}$, $A^r = \left(b^{\log_b A}\right)^r = b^{r\log_b A}$

Taking the log of both sides, $\log_b\left(A^r\right) = \log_b\left(b^{r\log_b A}\right)$

Utilizing the inverse property on the right side yields the result: $\log_b\left(A^r\right) = r\log_b A$

Example 8

Rewrite $\log_3(25)$ using the exponent property for logs.

Since $25 = 5^2$,
$\log_3(25) = \log_3\left(5^2\right) = 2\log_3(5)$

Example 9

Rewrite $4\ln(x)$ using the exponent property for logs.

Using the property in reverse, $4\ln(x) = \ln\left(x^4\right)$.

Try it Now

3. Rewrite using the exponent property for logs: $\ln\left(\dfrac{1}{x^2}\right)$.

The exponent property allows us to find a method for changing the base of a logarithmic expression.

Properties of Logs: Change of Base

$$\log_b(A) = \frac{\log_c(A)}{\log_c(b)}$$

Proof:

Let $\log_b(A) = x$.

Rewriting as an exponential gives $b^x = A$.

Taking the log base c of both sides of this equation gives $\log_c b^x = \log_c A$,

Now utilizing the exponent property for logs on the left side, $x\log_c b = \log_c A$

Dividing, we obtain $x = \dfrac{\log_c A}{\log_c b}$. Replacing our original expression for x,

$$\log_b A = \frac{\log_c A}{\log_c b}$$

With this change of base formula, we can finally find a good decimal approximation to our question from the beginning of the section.

Example 10

Evaluate $\log_2(10)$ using the change of base formula.

According to the change of base formula, we can rewrite the log base 2 as a logarithm of any other base. Since our calculators can evaluate the natural log, we might choose to use the natural logarithm, which is the log base e:

$$\log_2 10 = \frac{\log_e 10}{\log_e 2} = \frac{\ln 10}{\ln 2}$$

Using our calculators to evaluate this,

$$\frac{\ln 10}{\ln 2} \approx \frac{2.30259}{0.69315} \approx 3.3219$$

This finally allows us to answer our original question – the population of flies we discussed at the beginning of the section will take 3.32 weeks to grow to 500.

Example 11

Evaluate $\log_5(100)$ using the change of base formula.

We can rewrite this expression using any other base. If our calculators are able to evaluate the common logarithm, we could rewrite using the common log, base 10.

$$\log_5(100) = \frac{\log_{10} 100}{\log_{10} 5} \approx \frac{2}{0.69897} = 2.861$$

While we can solve the basic exponential equation $2^x = 10$ by rewriting in logarithmic form and then using the change of base formula to evaluate the logarithm, the proof of the change of base formula illuminates an alternative approach to solving exponential equations.

Solving exponential equations:

1. Isolate the exponential expressions when possible
2. Take the logarithm of both sides
3. Utilize the exponent property for logarithms to pull the variable out of the exponent
4. Use algebra to solve for the variable.

Example 12

Solve $2^x = 10$ for x.

Using this alternative approach, rather than rewrite this exponential into logarithmic form, we will take the logarithm of both sides of the equation. Since we often wish to evaluate the result to a decimal answer, we will usually utilize either the common log or natural log. For this example, we'll use the natural log:

$\ln(2^x) = \ln(10)$ Utilizing the exponent property for logs,

$x\ln(2) = \ln(10)$ Now dividing by ln(2),

$x = \dfrac{\ln(10)}{\ln(2)} \approx 3.3219$

Notice that this result matches the result we found using the change of base formula.

Example 13

In the first section, we predicted the population (in billions) of India t years after 2008 by using the function $f(t) = 1.14(1 + 0.0134)^t$. If the population continues following this trend, when will the population reach 2 billion?

We need to solve for time t so that $f(t) = 2$.

$2 = 1.14(1.0134)^t$ Divide by 1.14 to isolate the exponential expression

$\dfrac{2}{1.14} = 1.0134^t$ Take the logarithm of both sides of the equation

$\ln\left(\dfrac{2}{1.14}\right) = \ln\left(1.0134^t\right)$ Apply the exponent property on the right side

$\ln\left(\dfrac{2}{1.14}\right) = t\ln(1.0134)$ Divide both sides by ln(1.0134)

$t = \dfrac{\ln\left(\dfrac{2}{1.14}\right)}{\ln(1.0134)} \approx 42.23$ years

If this growth rate continues, the model predicts the population of India will reach 2 billion about 42 years after 2008, or approximately in the year 2050.

Try it Now

4. Solve $5(0.93)^x = 10$.

Example 14

Solve $5(1.07)^{3t} = 2$

To start, we want to isolate the exponential part of the expression, the $(1.07)^{3t}$, so it is alone on one side of the equation. Then we can use the log to solve the equation. We can use any base log; this time we'll use the common log.

$$5(1.07)^{3t} = 2$$ Divide both sides by 5 to isolate the exponential

$$(1.07)^{3t} = \frac{2}{5}$$ Take the log of both sides.

$$\log\left((1.07)^{3t}\right) = \log\left(\frac{2}{5}\right)$$ Use the exponent property for logs

$$3t \log(1.07) = \log\left(\frac{2}{5}\right)$$ Divide by $3\log(1.07)$ on both sides

$$\frac{3t \log(1.07)}{3\log(1.07)} = \frac{\log\left(\frac{2}{5}\right)}{3\log(1.07)}$$ Simplify and evaluate

$$t = \frac{\log\left(\frac{2}{5}\right)}{3\log(1.07)} \approx -4.5143$$

Note that when entering that expression on your calculator, be sure to put parentheses around the whole denominator to ensure the proper order of operations:
`log(2/5)/(3*log(1.07))`

In addition to solving exponential equations, logarithmic expressions are common in many physical situations.

Example 15

In chemistry, pH is a measure of the acidity or basicity of a liquid. The pH is related to the concentration of hydrogen ions, $[H^+]$, measured in moles per liter, by the equation $pH = -\log\left(\left[H^+\right]\right)$.

If a liquid has concentration of 0.0001 moles per liber, determine the pH.
Determine the hydrogen ion concentration of a liquid with pH of 7.

To answer the first question, we evaluate the expression $-\log(0.0001)$. While we could use our calculators for this, we do not really need them here, since we can use the inverse property of logs:
$$-\log(0.0001) = -\log\left(10^{-4}\right) = -(-4) = 4$$

To answer the second question, we need to solve the equation $7 = -\log\left(\left[H^+\right]\right)$. Begin by isolating the logarithm on one side of the equation by multiplying both sides by -1: $-7 = \log\left(\left[H^+\right]\right)$. Rewriting into exponential form yields the answer:
$$\left[H^+\right] = 10^{-7} = 0.0000001 \text{ moles per liter.}$$

Logarithms also provide us a mechanism for finding continuous growth models for exponential growth given two data points.

Example 15

A population grows from 100 to 130 in 2 weeks. Find the continuous growth rate.

Measuring t in weeks, we are looking for an equation $P(t) = ae^{rt}$ so that $P(0) = 100$ and $P(2) = 130$. Using the first pair of values,
$100 = ae^{r \cdot 0}$, so $a = 100$.

Using the second pair of values,

$130 = 100e^{r \cdot 2}$ Divide by 100

$\dfrac{130}{100} = e^{r2}$ Take the natural log of both sides

$\ln(1.3) = \ln\left(e^{r2}\right)$ Use the inverse property of logs

$\ln(1.3) = 2r$

$r = \dfrac{\ln(1.3)}{2} \approx 0.1312$

This population is growing at a continuous rate of 13.12% per week.

In general, we can relate the standard form of an exponential with the continuous growth form by noting (using k to represent the continuous growth rate to avoid the confusion of using r in two different ways in the same formula):

$a(1+r)^x = ae^{kx}$

$(1+r)^x = e^{kx}$

$1+r = e^k$

Converting Between Periodic to Continuous Growth Rate

In the equation $f(x) = a(1+r)^x$, r is the **periodic growth rate**, the percent growth each time period (weekly growth, annual growth, etc.).

In the equation $f(x) = ae^{kx}$, k is the **continuous growth rate**.

You can convert between these using: $1+r = e^k$.

Remember that the continuous growth rate k represents the nominal growth rate before accounting for the effects of continuous compounding, while r represents the actual percent increase in one time unit (one week, one year, etc.).

Example 16

A company's sales can be modeled by the function $S(t) = 5000e^{0.12t}$, with t measured in years. Find the annual growth rate.

Noting that $1 + r = e^k$, then $r = e^{0.12} - 1 = 0.1275$, so the annual growth rate is 12.75%. The sales function could also be written in the form $S(t) = 5000(1 + 0.1275)^t$.

Important Topics of this Section

The Logarithmic function as the inverse of the exponential function

Writing logarithmic & exponential expressions

Properties of logs

   Inverse properties

   Exponential properties

   Change of base

Common log

Natural log

Solving exponential equations

Converting between periodic and continuous growth rate.

Try it Now Answers

1. $\log_4(16) = 2 = \log_4 4^2 = 2\log_4 4$

2. $\log(1000000) = \log(10^6) = 6$

3. $\ln\left(\dfrac{1}{x^2}\right) = \ln(x^{-2}) = -2\ln(x)$

4. $5(0.93)^x = 10$

   $(0.93)^x = 2$

   $\ln(0.93^x) = \ln(2)$

   $x\ln(0.93) = \ln(2)$

   $\dfrac{\ln(2)}{\ln(0.93)} \approx -9.5513$

Section 4.3 Exercises

Rewrite each equation in exponential form
1. $\log_4(q) = m$ 2. $\log_3(t) = k$ 3. $\log_a(b) = c$ 4. $\log_p(z) = u$

5. $\log(v) = t$ 6. $\log(r) = s$ 7. $\ln(w) = n$ 8. $\ln(x) = y$

Rewrite each equation in logarithmic form.
9. $4^x = y$ 10. $5^y = x$ 11. $c^d = k$ 12. $n^z = L$

13. $10^a = b$ 14. $10^p = v$ 15. $e^k = h$ 16. $e^y = x$

Solve for x.
17. $\log_3(x) = 2$ 18. $\log_4(x) = 3$ 19. $\log_2(x) = -3$ 20. $\log_5(x) = -1$

21. $\log(x) = 3$ 22. $\log(x) = 5$ 23. $\ln(x) = 2$ 24. $\ln(x) = -2$

Simplify each expression using logarithm properties.
25. $\log_5(25)$ 26. $\log_2(8)$ 27. $\log_3\left(\dfrac{1}{27}\right)$ 28. $\log_6\left(\dfrac{1}{36}\right)$

29. $\log_6\left(\sqrt{6}\right)$ 30. $\log_5\left(\sqrt[3]{5}\right)$ 31. $\log(10,000)$ 32. $\log(100)$

33. $\log(0.001)$ 34. $\log(0.00001)$ 35. $\ln\left(e^{-2}\right)$ 36. $\ln\left(e^3\right)$

Evaluate using your calculator.
37. $\log(0.04)$ 38. $\log(1045)$ 39. $\ln(15)$ 40. $\ln(0.02)$

Solve each equation for the variable.
41. $5^x = 14$ 42. $3^x = 23$ 43. $7^x = \dfrac{1}{15}$ 44. $3^x = \dfrac{1}{4}$

45. $e^{5x} = 17$ 46. $e^{3x} = 12$ 47. $3^{4x-5} = 38$ 48. $4^{2x-3} = 44$

49. $1000(1.03)^t = 5000$ 50. $200(1.06)^t = 550$

51. $3(1.04)^{3t} = 8$ 52. $2(1.08)^{4t} = 7$

53. $50e^{-0.12t} = 10$ 54. $10e^{-0.03t} = 4$

55. $10 - 8\left(\dfrac{1}{2}\right)^x = 5$ 56. $100 - 100\left(\dfrac{1}{4}\right)^x = 70$

Convert the equation into continuous growth form, $f(t) = ae^{kt}$.

57. $f(t) = 300(0.91)^t$

58. $f(t) = 120(0.07)^t$

59. $f(t) = 10(1.04)^t$

60. $f(t) = 1400(1.12)^t$

Convert the equation into annual growth form, $f(t) = ab^t$.

61. $f(t) = 150e^{0.06t}$

62. $f(t) = 100e^{0.12t}$

63. $f(t) = 50e^{-0.012t}$

64. $f(t) = 80e^{-0.85t}$

65. The population of Kenya was 39.8 million in 2009 and has been growing by about 2.6% each year. If this trend continues, when will the population exceed 45 million?

66. The population of Algeria was 34.9 million in 2009 and has been growing by about 1.5% each year. If this trend continues, when will the population exceed 45 million?

67. The population of Seattle grew from 563,374 in 2000 to 608,660 in 2010. If the population continues to grow exponentially at the same rate, when will the population exceed 1 million people?

68. The median household income (adjusted for inflation) in Seattle grew from $42,948 in 1990 to $45,736 in 2000. If it continues to grow exponentially at the same rate, when will median income exceed $50,000?

69. A scientist begins with 100 mg of a radioactive substance. After 4 hours, it has decayed to 80 mg. How long after the process began will it take to decay to 15 mg?

70. A scientist begins with 100 mg of a radioactive substance. After 6 days, it has decayed to 60 mg. How long after the process began will it take to decay to 10 mg?

71. If $1000 is invested in an account earning 3% compounded monthly, how long will it take the account to grow in value to $1500?

72. If $1000 is invested in an account earning 2% compounded quarterly, how long will it take the account to grow in value to $1300?

Section 4.4 Logarithmic Properties

In the previous section, we derived two important properties of logarithms, which allowed us to solve some basic exponential and logarithmic equations.

Properties of Logs

Inverse Properties:

$$\log_b\left(b^x\right) = x$$

$$b^{\log_b x} = x$$

Exponential Property:

$$\log_b\left(A^r\right) = r \log_b\left(A\right)$$

Change of Base:

$$\log_b\left(A\right) = \frac{\log_c\left(A\right)}{\log_c\left(b\right)}$$

While these properties allow us to solve a large number of problems, they are not sufficient to solve all problems involving exponential and logarithmic equations.

Properties of Logs

Sum of Logs Property:

$$\log_b\left(A\right) + \log_b\left(C\right) = \log_b\left(AC\right)$$

Difference of Logs Property:

$$\log_b\left(A\right) - \log_b\left(C\right) = \log_b\left(\frac{A}{C}\right)$$

It's just as important to know what properties logarithms do *not* satisfy as to memorize the valid properties listed above. In particular, the logarithm is not a linear function, which means that it does not distribute: $\log(A + B) \neq \log(A) + \log(B)$.

To help in this process we offer a proof to help solidify our new rules and show how they follow from properties you've already seen.

Let $a = \log_b(A)$ and $c = \log_b(C)$.

By definition of the logarithm, $b^a = A$ and $b^c = C$.

Using these expressions, $AC = b^a b^c$

Using exponent rules on the right, $AC = b^{a+c}$

Taking the log of both sides, and utilizing the inverse property of logs,

$\log_b(AC) = \log_b(b^{a+c}) = a + c$

Replacing a and c with their definition establishes the result

$\log_b(AC) = \log_b A + \log_b C$

The proof for the difference property is very similar.

With these properties, we can rewrite expressions involving multiple logs as a single log, or break an expression involving a single log into expressions involving multiple logs.

Example 1

Write $\log_3(5) + \log_3(8) - \log_3(2)$ as a single logarithm.

Using the sum of logs property on the first two terms,

$\log_3(5) + \log_3(8) = \log_3(5 \cdot 8) = \log_3(40)$

This reduces our original expression to $\log_3(40) - \log_3(2)$

Then using the difference of logs property,

$\log_3(40) - \log_3(2) = \log_3\left(\dfrac{40}{2}\right) = \log_3(20)$

Example 2

Evaluate $2\log(5) + \log(4)$ without a calculator by first rewriting as a single logarithm.

On the first term, we can use the exponent property of logs to write

$2\log(5) = \log(5^2) = \log(25)$

With the expression reduced to a sum of two logs, $\log(25) + \log(4)$, we can utilize the sum of logs property

$\log(25) + \log(4) = \log(4 \cdot 25) = \log(100)$

Since $100 = 10^2$, we can evaluate this log without a calculator:

$\log(100) = \log(10^2) = 2$

Try it Now

1. Without a calculator evaluate by first rewriting as a single logarithm:

$$\log_2(8) + \log_2(4)$$

Example 3

Rewrite $\ln\left(\dfrac{x^4 y}{7}\right)$ as a sum or difference of logs

First, noticing we have a quotient of two expressions, we can utilize the difference property of logs to write

$$\ln\left(\frac{x^4 y}{7}\right) = \ln\left(x^4 y\right) - \ln(7)$$

Then seeing the product in the first term, we use the sum property
$$\ln\left(x^4 y\right) - \ln(7) = \ln\left(x^4\right) + \ln(y) - \ln(7)$$

Finally, we could use the exponent property on the first term
$$\ln\left(x^4\right) + \ln(y) - \ln(7) = 4\ln(x) + \ln(y) - \ln(7)$$

Interestingly, solving exponential equations was not the reason logarithms were originally developed. Historically, up until the advent of calculators and computers, the power of logarithms was that these log properties reduced multiplication, division, roots, or powers to be evaluated using addition, subtraction, division and multiplication, respectively, which are much easier to compute without a calculator. Large books were published listing the logarithms of numbers, such as in the table to the right. To find the product of two numbers, the sum of log property was used. Suppose for example we didn't know the value of 2 times 3. Using the sum property of logs:

value	log(value)
1	0.0000000
2	0.3010300
3	0.4771213
4	0.6020600
5	0.6989700
6	0.7781513
7	0.8450980
8	0.9030900
9	0.9542425
10	1.0000000

$$\log(2 \cdot 3) = \log(2) + \log(3)$$

Using the log table,
$$\log(2 \cdot 3) = \log(2) + \log(3) = 0.3010300 + 0.4771213 = 0.7781513$$

We can then use the table again in reverse, looking for 0.7781513 as an output of the logarithm. From that we can determine:
$$\log(2 \cdot 3) = 0.7781513 = \log(6).$$

By using addition and the table of logs, we were able to determine $2 \cdot 3 = 6$.

Likewise, to compute a cube root like $\sqrt[3]{8}$

$$\log(\sqrt[3]{8}) = \log\left(8^{1/3}\right) = \frac{1}{3}\log(8) = \frac{1}{3}(0.9030900) = 0.3010300 = \log(2)$$

So $\sqrt[3]{8} = 2$.

Although these calculations are simple and insignificant, they illustrate the same idea that was used for hundreds of years as an efficient way to calculate the product, quotient, roots, and powers of large and complicated numbers, either using tables of logarithms or mechanical tools called slide rules.

These properties still have other practical applications for interpreting changes in exponential and logarithmic relationships.

Example 4

Recall that in chemistry, $pH = -\log\left(\left[H^+\right]\right)$. If the concentration of hydrogen ions in a liquid is doubled, what is the affect on pH?

Suppose C is the original concentration of hydrogen ions, and P is the original pH of the liquid, so $P = -\log(C)$. If the concentration is doubled, the new concentration is $2C$. Then the pH of the new liquid is
$$pH = -\log(2C)$$

Using the sum property of logs,
$$pH = -\log(2C) = -\left(\log(2) + \log(C)\right) = -\log(2) - \log(C)$$

Since $P = -\log(C)$, the new pH is
$$pH = P - \log(2) = P - 0.301$$

When the concentration of hydrogen ions is doubled, the pH decreases by 0.301.

Log properties in solving equations

The logarithm properties often arise when solving problems involving logarithms. First, we'll look at a simpler log equation.

Example 5

Solve $\log(2x - 6) = 3$.

To solve for x, we need to get it out from inside the log function. There are two ways we can approach this.

Method 1: Rewrite as an exponential.

Recall that since the common log is base 10, $\log(A) = B$ can be rewritten as the exponential $10^B = A$. Likewise, $\log(2x - 6) = 3$ can be rewritten in exponential form as $10^3 = 2x - 6$

Method 2: Exponentiate both sides.

If $A = B$, then $10^A = 10^B$. Using this idea, since $\log(2x - 6) = 3$, then $10^{\log(2x-6)} = 10^3$. Use the inverse property of logs to rewrite the left side and get $2x - 6 = 10^3$.

Using either method, we now need to solve $2x - 6 = 10^3$. Evaluate 10^3 to get

$2x - 6 = 1000$ Add 6 to both sides
$2x = 1006$ Divide both sides by 2
$x = 503$

Occasionally the solving process will result in extraneous solutions – answers that are outside the domain of the original equation. In this case, our answer looks fine.

Example 6

Solve $\log(50x + 25) - \log(x) = 2$.

In order to rewrite in exponential form, we need a single logarithmic expression on the left side of the equation. Using the difference property of logs, we can rewrite the left side:

$$\log\left(\frac{50x + 25}{x}\right) = 2$$

Rewriting in exponential form reduces this to an algebraic equation:

$\dfrac{50x + 25}{x} = 10^2 = 100$ Multiply both sides by x

$50x + 25 = 100x$ Combine like terms

$25 = 50x$ Divide by 50

$x = \dfrac{25}{50} = \dfrac{1}{2}$

Checking this answer in the original equation, we can verify there are no domain issues, and this answer is correct.

Try it Now

2. Solve $\log(x^2 - 4) = 1 + \log(x + 2)$.

Example 7

Solve $\ln(x+2)+\ln(x+1) = \ln(4x+14)$.

$\ln(x+2)+\ln(x+1) = \ln(4x+14)$ Use the sum of logs property on the right

$\ln((x+2)(x+1)) = \ln(4x+14)$ Expand

$\ln(x^2+3x+2) = \ln(4x+14)$

We have a log on both side of the equation this time. Rewriting in exponential form would be tricky, so instead we can exponentiate both sides.

$e^{\ln(x^2+3x+2)} = e^{\ln(4x+13)}$ Use the inverse property of logs

$x^2+3x+2 = 4x+14$ Move terms to one side

$x^2-x-12 = 0$ Factor

$(x+4)(x-3) = 0$

$x = -4$ or $x = 3$.

Checking our answers, notice that evaluating the original equation at $x = -4$ would result in us evaluating $\ln(-2)$, which is undefined. That answer is outside the domain of the original equation, so it is an extraneous solution and we discard it. There is one solution: $x = 3$.

More complex exponential equations can often be solved in more than one way. In the following example, we will solve the same problem in two ways – one using logarithm properties, and the other using exponential properties.

Example 8a

In 2008, the population of Kenya was approximately 38.8 million, and was growing by 2.64% each year, while the population of Sudan was approximately 41.3 million and growing by 2.24% each year[2]. If these trends continue, when will the population of Kenya match that of Sudan?

We start by writing an equation for each population in terms of t, the number of years after 2008.

$Kenya(t) = 38.8(1+0.0264)^t$

$Sudan(t) = 41.3(1+0.0224)^t$

To find when the populations will be equal, we can set the equations equal

$38.8(1.0264)^t = 41.3(1.0224)^t$

---

[2] World Bank, World Development Indicators, as reported on http://www.google.com/publicdata, retrieved August 24, 2010

For our first approach, we take the log of both sides of the equation.

$$\log\left(38.8(1.0264)^t\right) = \log\left(41.3(1.0224)^t\right)$$

Utilizing the sum property of logs, we can rewrite each side,

$$\log(38.8) + \log\left(1.0264^t\right) = \log(41.3) + \log\left(1.0224^t\right)$$

Then utilizing the exponent property, we can pull the variables out of the exponent

$$\log(38.8) + t\log\left(1.0264\right) = \log(41.3) + t\log\left(1.0224\right)$$

Moving all the terms involving t to one side of the equation and the rest of the terms to the other side,

$$t\log\left(1.0264\right) - t\log\left(1.0224\right) = \log(41.3) - \log(38.8)$$

Factoring out the t on the left,

$$t\left(\log\left(1.0264\right) - \log\left(1.0224\right)\right) = \log(41.3) - \log(38.8)$$

Dividing to solve for t

$$t = \frac{\log(41.3) - \log(38.8)}{\log\left(1.0264\right) - \log\left(1.0224\right)} \approx 15.991 \text{ years until the populations will be equal.}$$

Example 8b

Solve the problem above by rewriting before taking the log.

Starting at the equation

$$38.8(1.0264)^t = 41.3(1.0224)^t$$

Divide to move the exponential terms to one side of the equation and the constants to the other side

$$\frac{1.0264^t}{1.0224^t} = \frac{41.3}{38.8}$$

Using exponent rules to group on the left,

$$\left(\frac{1.0264}{1.0224}\right)^t = \frac{41.3}{38.8}$$

Taking the log of both sides

$$\log\left(\left(\frac{1.0264}{1.0224}\right)^t\right) = \log\left(\frac{41.3}{38.8}\right)$$

Utilizing the exponent property on the left,

$$t \log\left(\frac{1.0264}{1.0224}\right) = \log\left(\frac{41.3}{38.8}\right)$$

Dividing gives

$$t = \frac{\log\left(\dfrac{41.3}{38.8}\right)}{\log\left(\dfrac{1.0264}{1.0224}\right)} \approx 15.991 \text{ years}$$

While the answer does not immediately appear identical to that produced using the previous method, note that by using the difference property of logs, the answer could be rewritten:

$$t = \frac{\log\left(\dfrac{41.3}{38.8}\right)}{\log\left(\dfrac{1.0264}{1.0224}\right)} = \frac{\log(41.3) - \log(38.8)}{\log(1.0264) - \log(1.0224)}$$

While both methods work equally well, it often requires fewer steps to utilize algebra before taking logs, rather than relying solely on log properties.

Try it Now
3. Tank A contains 10 liters of water, and 35% of the water evaporates each week. Tank B contains 30 liters of water, and 50% of the water evaporates each week. In how many weeks will the tanks contain the same amount of water?

Important Topics of this Section

Inverse

Exponential

Change of base

Sum of logs property

Difference of logs property

Solving equations using log rules

Try it Now Answers

1. $\log_2(8 \cdot 4) = \log_2(32) = \log_2(2^5) = 5$

2. $\log(x^2 - 4) = 1 + \log(x + 2)$ Move both logs to one side

    $\log(x^2 - 4) - \log(x + 2) = 1$ Use the difference property of logs

    $\log\left(\dfrac{x^2 - 4}{x + 2}\right) = 1$ Factor

    $\log\left(\dfrac{(x + 2)(x - 2)}{x + 2}\right) = 1$ Simplify

    $\log(x - 2) = 1$ Rewrite as an exponential

    $10^1 = x - 2$ Add 2 to both sides

    $x = 12$

3. Tank A: $A(t) = 10(1 - 0.35)^t$. Tank B: $B(t) = 30(1 - 0.50)^t$

    Solving $A(t) = B(t)$,

    $10(0.65)^t = 30(0.5)^t$ Using the method from Example 8b

    $\dfrac{(0.65)^t}{(0.5)^t} = \dfrac{30}{10}$ Regroup

    $\left(\dfrac{0.65}{0.5}\right)^t = 3$ Simplify

    $(1.3)^t = 3$ Take the log of both sides

    $\log\left((1.3)^t\right) = \log(3)$ Use the exponent property of logs

    $t \log(1.3) = \log(3)$ Divide and evaluate

    $t = \dfrac{\log(3)}{\log(1.3)} \approx 4.1874$ weeks

Section 4.4 Exercises

Simplify to a single logarithm, using logarithm properties.

1. $\log_3(28) - \log_3(7)$ 2. $\log_3(32) - \log_3(4)$

3. $-\log_3\left(\dfrac{1}{7}\right)$ 4. $-\log_4\left(\dfrac{1}{5}\right)$

5. $\log_3\left(\dfrac{1}{10}\right) + \log_3(50)$ 6. $\log_4(3) + \log_4(7)$

7. $\dfrac{1}{3}\log_7(8)$ 8. $\dfrac{1}{2}\log_5(36)$

9. $\log(2x^4) + \log(3x^5)$ 10. $\ln(4x^2) + \ln(3x^3)$

11. $\ln(6x^9) - \ln(3x^2)$ 12. $\log(12x^4) - \log(4x)$

13. $2\log(x) + 3\log(x+1)$ 14. $3\log(x) + 2\log(x^2)$

15. $\log(x) - \dfrac{1}{2}\log(y) + 3\log(z)$ 16. $2\log(x) + \dfrac{1}{3}\log(y) - \log(z)$

Use logarithm properties to expand each expression.

17. $\log\left(\dfrac{x^{15}y^{13}}{z^{19}}\right)$ 18. $\log\left(\dfrac{a^2b^3}{c^5}\right)$

19. $\ln\left(\dfrac{a^{-2}}{b^{-4}c^5}\right)$ 20. $\ln\left(\dfrac{a^{-2}b^3}{c^{-5}}\right)$

21. $\log\left(\sqrt{x^3y^{-4}}\right)$ 22. $\log\left(\sqrt{x^{-3}y^2}\right)$

23. $\ln\left(y\sqrt{\dfrac{y}{1-y}}\right)$ 24. $\ln\left(\dfrac{x}{\sqrt{1-x^2}}\right)$

25. $\log\left(x^2y^3\sqrt[3]{x^2y^5}\right)$ 26. $\log\left(x^3y^4\sqrt[7]{x^3y^9}\right)$

Solve each equation for the variable.

27. $4^{4x-7} = 3^{9x-6}$

28. $2^{2x-5} = 7^{3x-7}$

29. $17(1.14)^x = 19(1.16)^x$

30. $20(1.07)^x = 8(1.13)^x$

31. $5e^{0.12t} = 10e^{0.08t}$

32. $3e^{0.09t} = e^{0.14t}$

33. $\log_2(7x+6) = 3$

34. $\log_3(2x+4) = 2$

35. $2\ln(3x) + 3 = 1$

36. $4\ln(5x) + 5 = 2$

37. $\log(x^3) = 2$

38. $\log(x^5) = 3$

39. $\log(x) + \log(x+3) = 3$

40. $\log(x+4) + \log(x) = 9$

41. $\log(x+4) - \log(x+3) = 1$

42. $\log(x+5) - \log(x+2) = 2$

43. $\log_6(x^2) - \log_6(x+1) = 1$

44. $\log_3(x^2) - \log_3(x+2) = 5$

45. $\log(x+12) = \log(x) + \log(12)$

46. $\log(x+15) = \log(x) + \log(15)$

47. $\ln(x) + \ln(x-3) = \ln(7x)$

48. $\ln(x) + \ln(x-6) = \ln(6x)$

Section 4.5 Graphs of Logarithmic Functions

Recall that the exponential function $f(x) = 2^x$ produces this table of values

x	-3	-2	-1	0	1	2	3
$f(x)$	$\dfrac{1}{8}$	$\dfrac{1}{4}$	$\dfrac{1}{2}$	1	2	4	8

Since the logarithmic function is an inverse of the exponential, $g(x) = \log_2(x)$ produces the table of values

x	$\dfrac{1}{8}$	$\dfrac{1}{4}$	$\dfrac{1}{2}$	1	2	4	8
$g(x)$	-3	-2	-1	0	1	2	3

In this second table, notice that
1) As the input increases, the output increases.
2) As input increases, the output increases more slowly.
3) Since the exponential function only outputs positive values, the logarithm can only accept positive values as inputs, so the domain of the log function is $(0, \infty)$.
4) Since the exponential function can accept all real numbers as inputs, the logarithm can output any real number, so the range is all real numbers or $(-\infty, \infty)$.

Sketching the graph, notice that as the input approaches zero from the right, the output of the function grows very large in the negative direction, indicating a vertical asymptote at $x = 0$.
In symbolic notation we write
as $x \to 0^+, f(x) \to -\infty$, and as $x \to \infty, f(x) \to \infty$

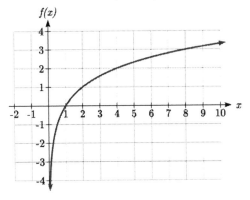

Graphical Features of the Logarithm

Graphically, in the function $g(x) = \log_b(x)$

The graph has a horizontal intercept at (1, 0)

The graph has a vertical asymptote at $x = 0$

The graph is increasing and concave down

The domain of the function is $x > 0$, or $(0, \infty)$

The range of the function is all real numbers, or $(-\infty, \infty)$

When sketching a general logarithm with base b, it can be helpful to remember that the graph will pass through the points (1, 0) and (b, 1).

To get a feeling for how the base affects the shape of the graph, examine the graphs below.

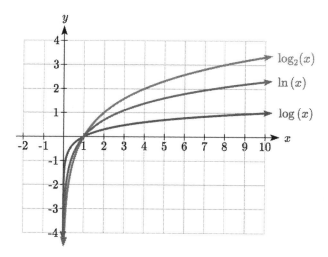

Notice that the larger the base, the slower the graph grows. For example, the common log graph, while it grows without bound, it does so very slowly. For example, to reach an output of 8, the input must be 100,000,000.

Another important observation made was the domain of the logarithm. Like the reciprocal and square root functions, the logarithm has a restricted domain which must be considered when finding the domain of a composition involving a log.

Example 1

Find the domain of the function $f(x) = \log(5 - 2x)$

The logarithm is only defined with the input is positive, so this function will only be defined when $5 - 2x > 0$. Solving this inequality,

$- 2x > -5$

$x < \dfrac{5}{2}$

The domain of this function is $x < \dfrac{5}{2}$, or in interval notation, $\left(-\infty, \dfrac{5}{2}\right)$

Try it Now

1. Find the domain of the function $f(x) = \log(x - 5) + 2$; before solving this as an inequality, consider how the function has been transformed.

Transformations of the Logarithmic Function

Transformations can be applied to a logarithmic function using the basic transformation techniques, but as with exponential functions, several transformations result in interesting relationships.

First recall the change of base property tells us that $\log_b x = \dfrac{\log_c x}{\log_c b} = \dfrac{1}{\log_c b}\log_c x$

From this, we can see that $\log_b x$ is a vertical stretch or compression of the graph of the $\log_c x$ graph. This tells us that a vertical stretch or compression is equivalent to a change of base. For this reason, we typically represent all graphs of logarithmic functions in terms of the common or natural log functions.

Next, consider the effect of a horizontal compression on the graph of a logarithmic function. Considering $f(x) = \log(cx)$, we can use the sum property to see
$f(x) = \log(cx) = \log(c) + \log(x)$

Since $\log(c)$ is a constant, the effect of a horizontal compression is the same as the effect of a vertical shift.

Example 2

Sketch $f(x) = \ln(x)$ and $g(x) = \ln(x) + 2$.

Graphing these,

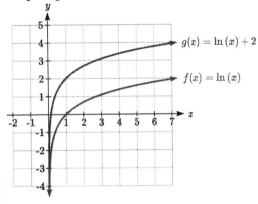

Note that this vertical shift could also be written as a horizontal compression, since
$g(x) = \ln(x) + 2 = \ln(x) + \ln(e^2) = \ln(e^2 x)$.

While a horizontal stretch or compression can be written as a vertical shift, a horizontal reflection is unique and separate from vertical shifting.

Finally, we will consider the effect of a horizontal shift on the graph of a logarithm.

Example 3

Sketch a graph of $f(x) = \ln(x+2)$.

This is a horizontal shift to the left by 2 units. Notice that none of our logarithm rules allow us rewrite this in another form, so the effect of this transformation is unique. Shifting the graph,

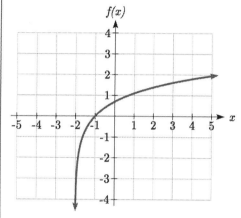

Notice that due to the horizontal shift, the vertical asymptote shifted to $x = -2$, and the domain shifted to $(-2, \infty)$.

Combining these transformations,

Example 4

Sketch a graph of $f(x) = 5\log(-x+2)$.

Factoring the inside as $f(x) = 5\log(-(x-2))$ reveals that this graph is that of the common logarithm, horizontally reflected, vertically stretched by a factor of 5, and shifted to the right by 2 units.

The vertical asymptote will be shifted to $x = 2$, and the graph will have domain $(\infty, 2)$. A rough sketch can be created by using the vertical asymptote along with a couple points on the graph, such as
$f(1) = 5\log(-1+2) = 5\log(1) = 0$
$f(-8) = 5\log(-(-8)+2) = 5\log(10) = 5$

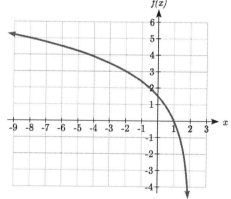

Try it Now
2. Sketch a graph of the function $f(x) = -3\log(x-2)+1$.

Transformations of Logs

Any transformed logarithmic function can be written in the form

$f(x) = a\log(x-b)+k$, or $f(x) = a\log\big(-(x-b)\big)+k$ if horizontally reflected,

where $x = b$ is the vertical asymptote.

Example 5

Find an equation for the logarithmic function graphed.

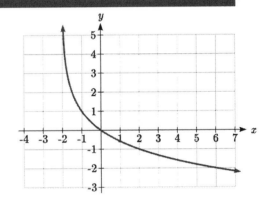

This graph has a vertical asymptote at $x = -2$ and has been vertically reflected. We do not know yet the vertical shift (equivalent to horizontal stretch) or the vertical stretch (equivalent to a change of base). We know so far that the equation will have form
$f(x) = -a\log(x+2)+k$

It appears the graph passes through the points (–1, 1) and (2, –1). Substituting in (–1, 1),
$1 = -a\log(-1+2)+k$

$1 = -a\log(1)+k$

$1 = k$

Next, substituting in (2, –1),
$-1 = -a\log(2+2)+1$

$-2 = -a\log(4)$

$a = \dfrac{2}{\log(4)}$

This gives us the equation $f(x) = -\dfrac{2}{\log(4)}\log(x+2)+1$.

This could also be written as $f(x) = -2\log_4(x+2)+1$.

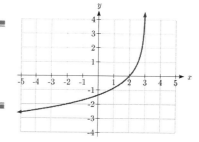

Try it Now

3. Write an equation for the function graphed here.

Flashback

4. Write the domain and range of the function graphed in Example 5, and describe its long run behavior.

Important Topics of this Section
Graph of the logarithmic function (domain and range)
Transformation of logarithmic functions
Creating graphs from equations
Creating equations from graphs

Try it Now and Flashback Answers

1. Domain: $\{x \mid x > 5\}$

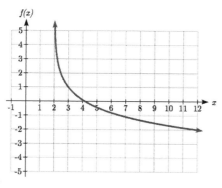

2.

3. The graph is horizontally reflected and has a vertical asymptote at $x = 3$, giving form $f(x) = a \log(-(x-3)) + k$. Substituting in the point (2,0) gives $0 = a \log(-(2-3)) + k$, simplifying to $k = 0$. Substituting in (-2,-2), $-2 = a \log(-(-2-3))$, so $\dfrac{-2}{\log(5)} = a$.

The equation is $f(x) = \dfrac{-2}{\log(5)} \log(-(x-3))$ or $f(x) = -2 \log_5 (-(x-3))$.

4. Domain: $\{x \mid x > -2\}$, Range: all real numbers; As $x \to -2^+$, $f(x) \to \infty$ and as $x \to \infty, f(x) \to -\infty$.

Section 4.5 Exercises

For each function, find the domain and the vertical asymptote.

1. $f(x) = \log(x-5)$

2. $f(x) = \log(x+2)$

3. $f(x) = \ln(3-x)$

4. $f(x) = \ln(5-x)$

5. $f(x) = \log(3x+1)$

6. $f(x) = \log(2x+5)$

7. $f(x) = 3\log(-x) + 2$

8. $f(x) = 2\log(-x) + 1$

Sketch a graph of each pair of functions.

9. $f(x) = \log(x), g(x) = \ln(x)$

10. $f(x) = \log_2(x), g(x) = \log_4(x)$

Sketch each transformation.

11. $f(x) = 2\log(x)$

12. $f(x) = 3\ln(x)$

13. $f(x) = \ln(-x)$

14. $f(x) = -\log(x)$

15. $f(x) = \log_2(x+2)$

16. $f(x) = \log_3(x+4)$

Find a formula for the transformed logarithm graph shown.

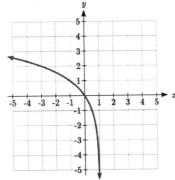

17.

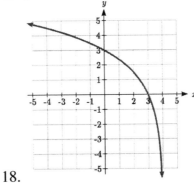

18.

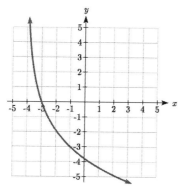

19.

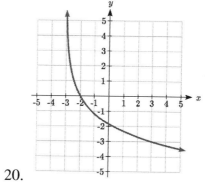

20.

Find a formula for the transformed logarithm graph shown.

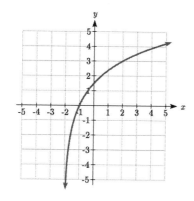

21.

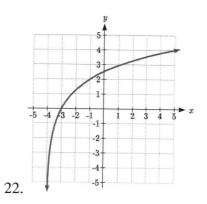

22.

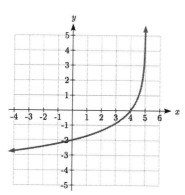

23.

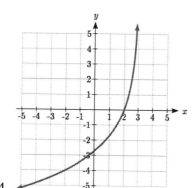

24.

Section 4.6 Exponential and Logarithmic Models

While we have explored some basic applications of exponential and logarithmic functions, in this section we explore some important applications in more depth.

Radioactive Decay

In an earlier section, we discussed radioactive decay – the idea that radioactive isotopes change over time. One of the common terms associated with radioactive decay is half-life.

Half Life

The **half-life** of a radioactive isotope is the time it takes for half the substance to decay.

Given the basic exponential growth/decay equation $h(t) = ab^t$, half-life can be found by solving for when half the original amount remains; by solving $\frac{1}{2}a = a(b)^t$, or more simply $\frac{1}{2} = b^t$. Notice how the initial amount is irrelevant when solving for half-life.

Example 1

Bismuth-210 is an isotope that decays by about 13% each day. What is the half-life of Bismuth-210?

We were not given a starting quantity, so we could either make up a value or use an unknown constant to represent the starting amount. To show that starting quantity does not affect the result, let us denote the initial quantity by the constant a. Then the decay of Bismuth-210 can be described by the equation $Q(d) = a(0.87)^d$.

To find the half-life, we want to determine when the remaining quantity is half the original: $\frac{1}{2}a$. Solving,

$\frac{1}{2}a = a(0.87)^d$ Divide by a,

$\frac{1}{2} = 0.87^d$ Take the log of both sides

$\log\left(\frac{1}{2}\right) = \log\left(0.87^d\right)$ Use the exponent property of logs

$$\log\left(\frac{1}{2}\right) = d \log(0.87) \qquad \text{Divide to solve for } d$$

$$d = \frac{\log\left(\frac{1}{2}\right)}{\log(0.87)} \approx 4.977 \text{ days}$$

This tells us that the half-life of Bismuth-210 is approximately 5 days.

Example 2

Cesium-137 has a half-life of about 30 years. If you begin with 200mg of cesium-137, how much will remain after 30 years? 60 years? 90 years?

Since the half-life is 30 years, after 30 years, half the original amount, 100mg, will remain.

After 60 years, another 30 years have passed, so during that second 30 years, another half of the substance will decay, leaving 50mg.

After 90 years, another 30 years have passed, so another half of the substance will decay, leaving 25mg.

Example 3

Cesium-137 has a half-life of about 30 years. Find the annual decay rate.

Since we are looking for an annual decay rate, we will use an equation of the form $Q(t) = a(1+r)^t$. We know that after 30 years, half the original amount will remain. Using this information

$$\frac{1}{2}a = a(1+r)^{30} \qquad \text{Dividing by } a$$

$$\frac{1}{2} = (1+r)^{30} \qquad \text{Taking the } 30^{\text{th}} \text{ root of both sides}$$

$$\sqrt[30]{\frac{1}{2}} = 1+r \qquad \text{Subtracting one from both sides,}$$

$$r = \sqrt[30]{\frac{1}{2}} - 1 \approx -0.02284$$

This tells us cesium-137 is decaying at an annual rate of 2.284% per year.

1. Chlorine-36 is eliminated from the body with a biological half-life of 10 days[3]. Find the daily decay rate.

Example 4

Carbon-14 is a radioactive isotope that is present in organic materials, and is commonly used for dating historical artifacts. Carbon-14 has a half-life of 5730 years. If a bone fragment is found that contains 20% of its original carbon-14, how old is the bone?

To find how old the bone is, we first will need to find an equation for the decay of the carbon-14. We could either use a continuous or annual decay formula, but opt to use the continuous decay formula since it is more common in scientific texts. The half life tells us that after 5730 years, half the original substance remains. Solving for the rate,

$\dfrac{1}{2}a = ae^{r5730}$ Dividing by a

$\dfrac{1}{2} = e^{r5730}$ Taking the natural log of both sides

$\ln\left(\dfrac{1}{2}\right) = \ln\left(e^{r5730}\right)$ Use the inverse property of logs on the right side

$\ln\left(\dfrac{1}{2}\right) = 5730r$ Divide by 5730

$r = \dfrac{\ln\left(\dfrac{1}{2}\right)}{5730} \approx -0.000121$

Now we know the decay will follow the equation $Q(t) = ae^{-0.000121\,t}$. To find how old the bone fragment is that contains 20% of the original amount, we solve for t so that $Q(t) = 0.20a$.

$0.20a = ae^{-0.000121t}$

$0.20 = e^{-0.000121t}$

$\ln(0.20) = \ln\left(e^{-0.000121\,t}\right)$

$\ln(0.20) = -0.000121t$

$t = \dfrac{\ln(0.20)}{-0.000121} \approx 13301 \ \text{years}$

The bone fragment is about 13,300 years old.

---

[3] http://www.ead.anl.gov/pub/doc/chlorine.pdf

2. In Example 2, we learned that Cesium-137 has a half-life of about 30 years. If you begin with 200mg of cesium-137, will it take more or less than 230 years until only 1 milligram remains?

Doubling Time

For decaying quantities, we asked how long it takes for half the substance to decay. For growing quantities we might ask how long it takes for the quantity to double.

> **Doubling Time**
>
> The **doubling time** of a growing quantity is the time it takes for the quantity to double.

Given the basic exponential growth equation $h(t) = ab^t$, doubling time can be found by solving for when the original quantity has doubled; by solving $2a = a(b)^x$, or more simply $2 = b^x$. Like with decay, the initial amount is irrelevant when solving for doubling time.

Example 5

Cancer cells sometimes increase exponentially. If a cancerous growth contained 300 cells last month and 360 cells this month, how long will it take for the number of cancer cells to double?

Defining t to be time in months, with $t = 0$ corresponding to this month, we are given two pieces of data: this month, (0, 360), and last month, (-1, 300).

From this data, we can find an equation for the growth. Using the form $C(t) = ab^t$, we know immediately $a = 360$, giving $C(t) = 360b^t$. Substituting in (-1, 300),

$$300 = 360b^{-1}$$

$$300 = \frac{360}{b}$$

$$b = \frac{360}{300} = 1.2$$

This gives us the equation $C(t) = 360(1.2)^t$

To find the doubling time, we look for the time when we will have twice the original amount, so when $C(t) = 2a$.

$2a = a(1.2)^t$

$2 = (1.2)^t$

$\log(2) = \log(1.2^t)$

$\log(2) = t\log(1.2)$

$t = \dfrac{\log(2)}{\log(1.2)} \approx 3.802$ months for the number of cancer cells to double.

Example 6

Use of a new social networking website has been growing exponentially, with the number of new members doubling every 5 months. If the site currently has 120,000 users and this trend continues, how many users will the site have in 1 year?

We can use the doubling time to find a function that models the number of site users, and then use that equation to answer the question. While we could use an arbitrary a as we have before for the initial amount, in this case, we know the initial amount was 120,000.

If we use a continuous growth equation, it would look like $N(t) = 120e^{rt}$, measured in thousands of users after t months. Based on the doubling time, there would be 240 thousand users after 5 months. This allows us to solve for the continuous growth rate:

$240 = 120e^{r5}$

$2 = e^{r5}$

$\ln 2 = 5r$

$r = \dfrac{\ln 2}{5} \approx 0.1386$

Now that we have an equation, $N(t) = 120e^{0.1386t}$, we can predict the number of users after 12 months:

$N(12) = 120e^{0.1386(12)} = 633.140$ thousand users.

So after 1 year, we would expect the site to have around 633,140 users.

Try it Now

3. If tuition at a college is increasing by 6.6% each year, how many years will it take for tuition to double?

Newton's Law of Cooling

When a hot object is left in surrounding air that is at a lower temperature, the object's temperature will decrease exponentially, leveling off towards the surrounding air temperature. This "leveling off" will correspond to a horizontal asymptote in the graph of the temperature function. Unless the room temperature is zero, this will correspond to a vertical shift of the generic exponential decay function.

Newton's Law of Cooling

The temperature of an object, T, in surrounding air with temperature T_s will behave according to the formula

$$T(t) = ae^{kt} + T_s$$

Where

t is time

a is a constant determined by the initial temperature of the object

k is a constant, the continuous rate of cooling of the object

---

While an equation of the form $T(t) = ab^t + T_s$ could be used, the continuous growth form is more common.

Example 7

A cheesecake is taken out of the oven with an ideal internal temperature of 165 degrees Fahrenheit, and is placed into a 35 degree refrigerator. After 10 minutes, the cheesecake has cooled to 150 degrees. If you must wait until the cheesecake has cooled to 70 degrees before you eat it, how long will you have to wait?

Since the surrounding air temperature in the refrigerator is 35 degrees, the cheesecake's temperature will decay exponentially towards 35, following the equation
$$T(t) = ae^{kt} + 35$$

We know the initial temperature was 165, so $T(0) = 165$. Substituting in these values,

$165 = ae^{k0} + 35$

$165 = a + 35$

$a = 130$

We were given another pair of data, $T(10) = 150$, which we can use to solve for k

$150 = 130e^{k10} + 35$

$115 = 130e^{k10}$

$\dfrac{115}{130} = e^{10k}$

$\ln\left(\dfrac{115}{130}\right) = 10k$

$k = \dfrac{\ln\left(\dfrac{115}{130}\right)}{10} = -0.0123$

Together this gives us the equation for cooling: $T(t) = 130e^{-0.0123t} + 35$.

Now we can solve for the time it will take for the temperature to cool to 70 degrees.

$70 = 130e^{-0.0123t} + 35$

$35 = 130e^{-0.0123t}$

$\dfrac{35}{130} = e^{-0.0123t}$

$\ln\left(\dfrac{35}{130}\right) = -0.0123t$

$t = \dfrac{\ln\left(\dfrac{35}{130}\right)}{-0.0123} \approx 106.68$

It will take about 107 minutes, or one hour and 47 minutes, for the cheesecake to cool. Of course, if you like your cheesecake served chilled, you'd have to wait a bit longer.

Try it Now

4. A pitcher of water at 40 degrees Fahrenheit is placed into a 70 degree room. One hour later the temperature has risen to 45 degrees. How long will it take for the temperature to rise to 60 degrees?

Logarithmic Scales

For quantities that vary greatly in magnitude, a standard scale of measurement is not always effective, and utilizing logarithms can make the values more manageable. For example, if the average distances from the sun to the major bodies in our solar system are listed, you see they vary greatly.

Planet	Distance (millions of km)
Mercury	58
Venus	108
Earth	150
Mars	228
Jupiter	779
Saturn	1430
Uranus	2880
Neptune	4500

Placed on a linear scale – one with equally spaced values – these values get bunched up.

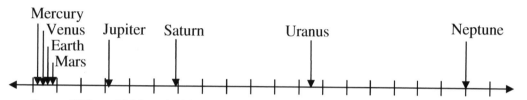

However, computing the logarithm of each value and plotting these new values on a number line results in a more manageable graph, and makes the relative distances more apparent.[4]

Planet	Distance (millions of km)	log(distance)
Mercury	58	1.76
Venus	108	2.03
Earth	150	2.18
Mars	228	2.36
Jupiter	779	2.89
Saturn	1430	3.16
Uranus	2880	3.46
Neptune	4500	3.65

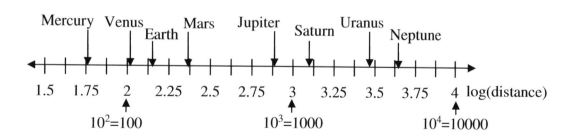

---

[4] It is interesting to note the large gap between Mars and Jupiter on the log number line. The asteroid belt is located there, which scientists believe is a planet that never formed because of the effects of the gravity of Jupiter.

Sometimes, as shown above, the scale on a logarithmic number line will show the log values, but more commonly the original values are listed as powers of 10, as shown below.

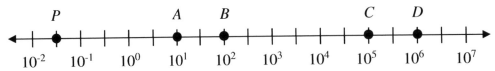

Example 8

Estimate the value of point P on the log scale above

The point P appears to be half way between -2 and -1 in log value, so if V is the value of this point,

$$\log(V) \approx -1.5 \qquad \text{Rewriting in exponential form,}$$

$$V \approx 10^{-1.5} = 0.0316$$

Example 9

Place the number 6000 on a logarithmic scale.

Since $\log(6000) \approx 3.8$, this point would belong on the log scale about here:

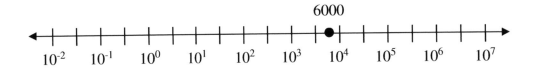

Try it Now

5. Plot the data in the table below on a logarithmic scale[5].

Source of Sound/Noise	Approximate Sound Pressure in μPa (micro Pascals)
Launching of the Space Shuttle	2,000,000,000
Full Symphony Orchestra	2,000,000
Diesel Freight Train at High Speed at 25 m	200,000
Normal Conversation	20,000
Soft Whispering at 2 m in Library	2,000
Unoccupied Broadcast Studio	200
Softest Sound a human can hear	20

---

[5] From http://www.epd.gov.hk/epd/noise_education/web/ENG_EPD_HTML/m1/intro_5.html, retrieved Oct 2, 2010

Notice that on the log scale above Example 8, the visual distance on the scale between points A and B and between C and D is the same. When looking at the values these points correspond to, notice B is ten times the value of A, and D is ten times the value of C. A visual *linear* difference between points corresponds to a *relative* (ratio) change between the corresponding values.

Logarithms are useful for showing these relative changes. For example, comparing $1,000,000 to $10,000, the first is 100 times larger than the second.

$$\frac{1,000,000}{10,000} = 100 = 10^2$$

Likewise, comparing $1000 to $10, the first is 100 times larger than the second.

$$\frac{1,000}{10} = 100 = 10^2$$

When one quantity is roughly ten times larger than another, we say it is one **order of magnitude** larger. In both cases described above, the first number was two orders of magnitude larger than the second.

Notice that the order of magnitude can be found as the common logarithm of the ratio of the quantities. On the log scale above, B is one order of magnitude larger than A, and D is one order of magnitude larger than C.

Orders of Magnitude

Given two values A and B, to determine how many **orders of magnitude** A is greater than B,

$$\text{Difference in orders of magnitude} = \log\left(\frac{A}{B}\right)$$

Example 10

On the log scale above Example 8, how many orders of magnitude larger is C than B?

The value B corresponds to $10^2 = 100$
The value C corresponds to $10^5 = 100,000$

The relative change is $\dfrac{100,000}{100} = 1000 = \dfrac{10^5}{10^2} = 10^3$. The log of this value is 3.

C is three orders of magnitude greater than B, which can be seen on the log scale by the visual difference between the points on the scale.

Try it Now
6. Using the table from Try it Now #5, what is the difference of order of magnitude between the softest sound a human can hear and the launching of the space shuttle?

Earthquakes

An example of a logarithmic scale is the Moment Magnitude Scale (MMS) used for earthquakes. This scale is commonly and mistakenly called the Richter Scale, which was a very similar scale succeeded by the MMS.

Moment Magnitude Scale

For an earthquake with seismic moment S, a measurement of earth movement, the MMS value, or magnitude of the earthquake, is

$$M = \frac{2}{3}\log\left(\frac{S}{S_0}\right)$$

Where $S_0 = 10^{16}$ is a baseline measure for the seismic moment.

Example 11

If one earthquake has a MMS magnitude of 6.0, and another has a magnitude of 8.0, how much more powerful (in terms of earth movement) is the second earthquake?

Since the first earthquake has magnitude 6.0, we can find the amount of earth movement for that quake, which we'll denote S_1. The value of S_0 is not particularly relevant, so we will not replace it with its value.

$$6.0 = \frac{2}{3}\log\left(\frac{S_1}{S_0}\right)$$

$$6.0\left(\frac{3}{2}\right) = \log\left(\frac{S_1}{S_0}\right)$$

$$9 = \log\left(\frac{S_1}{S_0}\right)$$

$$\frac{S_1}{S_0} = 10^9$$

$$S_1 = 10^9 S_0$$

This tells us the first earthquake has about 10^9 times more earth movement than the baseline measure.

Doing the same with the second earthquake, S_2, with a magnitude of 8.0,

$$8.0 = \frac{2}{3}\log\left(\frac{S_2}{S_0}\right)$$

$$S_2 = 10^{12} S_0$$

Comparing the earth movement of the second earthquake to the first,

$$\frac{S_2}{S_1} = \frac{10^{12} S_0}{10^9 S_0} = 10^3 = 1000$$

The second value's earth movement is 1000 times as large as the first earthquake.

Example 12

One earthquake has magnitude of 3.0. If a second earthquake has twice as much earth movement as the first earthquake, find the magnitude of the second quake.

Since the first quake has magnitude 3.0,

$$3.0 = \frac{2}{3}\log\left(\frac{S}{S_0}\right)$$

Solving for S,

$$3.0\left(\frac{3}{2}\right) = \log\left(\frac{S}{S_0}\right)$$

$$4.5 = \log\left(\frac{S}{S_0}\right)$$

$$10^{4.5} = \frac{S}{S_0}$$

$$S = 10^{4.5} S_0$$

Since the second earthquake has twice as much earth movement, for the second quake,
$$S = 2 \cdot 10^{4.5} S_0$$

Finding the magnitude,

$$M = \frac{2}{3}\log\left(\frac{2 \cdot 10^{4.5} S_0}{S_0}\right)$$

$$M = \frac{2}{3}\log\left(2 \cdot 10^{4.5}\right) \approx 3.201$$

The second earthquake with twice as much earth movement will have a magnitude of about 3.2.

In fact, using log properties, we could show that whenever the earth movement doubles, the magnitude will increase by about 0.201:

$$M = \frac{2}{3}\log\left(\frac{2S}{S_0}\right) = \frac{2}{3}\log\left(2 \cdot \frac{S}{S_0}\right)$$

$$M = \frac{2}{3}\left(\log(2) + \log\left(\frac{S}{S_0}\right)\right)$$

$$M = \frac{2}{3}\log(2) + \frac{2}{3}\log\left(\frac{S}{S_0}\right)$$

$$M = 0.201 + \frac{2}{3}\log\left(\frac{S}{S_0}\right)$$

This illustrates the most important feature of a log scale: that *multiplying* the quantity being considered will *add* to the scale value, and vice versa.

Important Topics of this Section
Radioactive decay
Half life
Doubling time
Newton's law of cooling
Logarithmic Scales
Orders of Magnitude
Moment Magnitude scale

Try it Now Answers

1. $r = \sqrt[10]{\dfrac{1}{2}} - 1 \approx -0.067$ or 6.7% is the daily rate of decay.

2. Less than 230 years, 229.3157 to be exact

3. Solving $a(1+0.066)^t = 2a$, it will take $t = \dfrac{\log(2)}{\log(1.066)} \approx 10.845$ years, or approximately 11 years, for tuition to double.

4. $T(t) = ae^{kt} + 70$. Substituting (0, 40), we find $a = -30$. Substituting (1, 45), we solve $45 = -30e^{k(1)} + 70$ to get $k = \ln\left(\dfrac{25}{30}\right) = -0.1823$.

   Solving $60 = -30e^{-0.1823t} + 70$ gives $t = \dfrac{\ln(1/3)}{-0.1823} = 6.026$ hours

5.

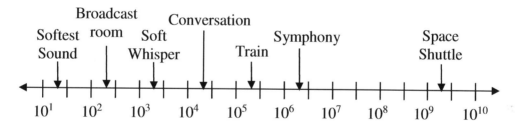

6. $\dfrac{2x10^9}{2x10^1} = 10^8$. The sound pressure in μPa created by launching the space shuttle is 8 orders of magnitude greater than the sound pressure in μPa created by the softest sound a human ear can hear.

Section 4.6 Exercises

1. You go to the doctor and he injects you with 13 milligrams of radioactive dye. After 12 minutes, 4.75 milligrams of dye remain in your system. To leave the doctor's office, you must pass through a radiation detector without sounding the alarm. If the detector will sound the alarm whenever more than 2 milligrams of the dye are in your system, how long will your visit to the doctor take, assuming you were given the dye as soon as you arrived and the amount of dye decays exponentially?

2. You take 200 milligrams of a headache medicine, and after 4 hours, 120 milligrams remain in your system. If the effects of the medicine wear off when less than 80 milligrams remain, when will you need to take a second dose, assuming the amount of medicine in your system decays exponentially?

3. The half-life of Radium-226 is 1590 years. If a sample initially contains 200 mg, how many milligrams will remain after 1000 years?

4. The half-life of Fermium-253 is 3 days. If a sample initially contains 100 mg, how many milligrams will remain after 1 week?

5. The half-life of Erbium-165 is 10.4 hours. After 24 hours a sample still contains 2 mg. What was the initial mass of the sample, and how much will remain after another 3 days?

6. The half-life of Nobelium-259 is 58 minutes. After 3 hours a sample still contains 10 mg. What was the initial mass of the sample, and how much will remain after another 8 hours?

7. A scientist begins with 250 grams of a radioactive substance. After 225 minutes, the sample has decayed to 32 grams. Find the half-life of this substance.

8. A scientist begins with 20 grams of a radioactive substance. After 7 days, the sample has decayed to 17 grams. Find the half-life of this substance.

9. A wooden artifact from an archeological dig contains 60 percent of the carbon-14 that is present in living trees. How long ago was the artifact made? (The half-life of carbon-14 is 5730 years.)

10. A wooden artifact from an archeological dig contains 15 percent of the carbon-14 that is present in living trees. How long ago was the artifact made? (The half-life of carbon-14 is 5730 years.)

11. A bacteria culture initially contains 1500 bacteria and doubles in size every half hour. Find the size of the population after: a) 2 hours   b) 100 minutes

12. A bacteria culture initially contains 2000 bacteria and doubles in size every half hour. Find the size of the population after: a) 3 hours   b) 80 minutes

13. The count of bacteria in a culture was 800 after 10 minutes and 1800 after 40 minutes.
    a. What was the initial size of the culture?
    b. Find the doubling time.
    c. Find the population after 105 minutes.
    d. When will the population reach 11000?

14. The count of bacteria in a culture was 600 after 20 minutes and 2000 after 35 minutes.
    a. What was the initial size of the culture?
    b. Find the doubling time.
    c. Find the population after 170 minutes.
    d. When will the population reach 12000?

15. Find the time required for an investment to double in value if invested in an account paying 3% compounded quarterly.

16. Find the time required for an investment to double in value if invested in an account paying 4% compounded monthly

17. The number of crystals that have formed after t hours is given by $n(t) = 20e^{0.013t}$. How long does it take the number of crystals to double?

18. The number of building permits in Pasco t years after 1992 roughly followed the equation $n(t) = 400e^{0.143t}$. What is the doubling time?

19. A turkey is pulled from the oven when the internal temperature is 165° Fahrenheit, and is allowed to cool in a 75° room. If the temperature of the turkey is 145° after half an hour,
    a. What will the temperature be after 50 minutes?
    b. How long will it take the turkey to cool to 110°?

20. A cup of coffee is poured at 190° Fahrenheit, and is allowed to cool in a 70° room. If the temperature of the coffee is 170° after half an hour,
    a. What will the temperature be after 70 minutes?
    b. How long will it take the coffee to cool to 120°?

21. The population of fish in a farm-stocked lake after t years could be modeled by the equation $P(t) = \dfrac{1000}{1 + 9e^{-0.6t}}$.
    a. Sketch a graph of this equation.
    b. What is the initial population of fish?
    c. What will the population be after 2 years?
    d. How long will it take for the population to reach 900?

22. The number of people in a town who have heard a rumor after t days can be modeled by the equation $N(t) = \dfrac{500}{1 + 49e^{-0.7t}}$.
    a. Sketch a graph of this equation.
    b. How many people started the rumor?
    c. How many people have heard the rumor after 3 days?
    d. How long will it take until 300 people have heard the rumor?

Find the value of the number shown on each logarithmic scale

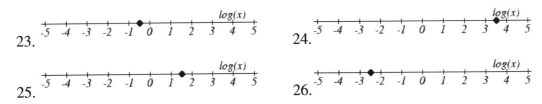

23.   24.

25.   26.

Plot each set of approximate values on a logarithmic scale.

27. Intensity of sounds: Whisper: $10^{-10} \ W/m^2$, Vacuum: $10^{-4} W/m^2$, Jet: $10^2 \ W/m^2$

28. Mass: Amoeba: $10^{-5} g$, Human: $10^5 g$, Statue of Liberty: $10^8 g$

29. The 1906 San Francisco earthquake had a magnitude of 7.9 on the MMS scale. Later there was an earthquake with magnitude 4.7 that caused only minor damage. How many times more intense was the San Francisco earthquake than the second one?

30. The 1906 San Francisco earthquake had a magnitude of 7.9 on the MMS scale. Later there was an earthquake with magnitude 6.5 that caused less damage. How many times more intense was the San Francisco earthquake than the second one?

31. One earthquake has magnitude 3.9 on the MMS scale. If a second earthquake has 750 times as much energy as the first, find the magnitude of the second quake.

32. One earthquake has magnitude 4.8 on the MMS scale. If a second earthquake has 1200 times as much energy as the first, find the magnitude of the second quake.

33. A colony of yeast cells is estimated to contain 10^6 cells at time $t = 0$. After collecting experimental data in the lab, you decide that the total population of cells at time t hours is given by the function $f(t) = 10^6 e^{0.495105t}$.   [UW]

    a. How many cells are present after one hour?
    b. How long does it take of the population to double?.
    c. Cherie, another member of your lab, looks at your notebook and says: "That formula is wrong, my calculations predict the formula for the number of yeast cells is given by the function. $f(t) = 10^6 (2.042727)^{0.693147t}$." Should you be worried by Cherie's remark?
    d. Anja, a third member of your lab working with the same yeast cells, took these two measurements: 7.246×10^6 cells after 4 hours; 16.504×10^6 cells after 6 hours. Should you be worried by Anja's results? If Anja's measurements are correct, does your model over estimate or under estimate the number of yeast cells at time t?

34. As light from the surface penetrates water, its intensity is diminished. In the clear waters of the Caribbean, the intensity is decreased by 15 percent for every 3 meters of depth. Thus, the intensity will have the form of a general exponential function. [UW]

    a. If the intensity of light at the water's surface is I_0, find a formula for $I(d)$, the intensity of light at a depth of d meters. Your formula should depend on I_0 and d.
    b. At what depth will the light intensity be decreased to 1% of its surface intensity?

35. Myoglobin and hemoglobin are oxygen-carrying molecules in the human body. Hemoglobin is found inside red blood cells, which flow from the lungs to the muscles through the bloodstream. Myoglobin is found in muscle cells. The function

$$Y = M(p) = \frac{p}{1+p}$$ calculates the fraction of myoglobin saturated with oxygen at a

given pressure p Torrs. For example, at a pressure of 1 Torr, $M(1) = 0.5$, which means half of the myoglobin (i.e. 50%) is oxygen saturated. (Note: More precisely, you need to use something called the "partial pressure", but the distinction is not important for

this problem.) Likewise, the function $Y = H(p) = \dfrac{p^{2.8}}{26^{2.8} + p^{2.8}}$ calculates the fraction

of hemoglobin saturated with oxygen at a given pressure p. [UW]

   a. The graphs of $M(p)$ and $H(p)$ are given here on the domain $0 \le p \le 100$; which is which?

   b. If the pressure in the lungs is 100 Torrs, what is the level of oxygen saturation of the hemoglobin in the lungs?

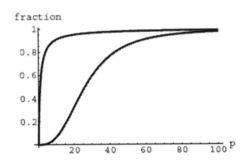

   c. The pressure in an active muscle is 20 Torrs. What is the level of oxygen saturation of myoglobin in an active muscle? What is the level of hemoglobin in an active muscle?

   d. Define the efficiency of oxygen transport at a given pressure p to be $M(p) - H(p)$. What is the oxygen transport efficiency at 20 Torrs? At 40 Torrs? At 60 Torrs? Sketch the graph of $M(p) - H(p)$; are there conditions under which transport efficiency is maximized (explain)?

36. The length of some fish are modeled by a von Bertalanffy growth function. For Pacific halibut, this function has the form $L(t) = 200\left(1 - 0.957e^{-0.18t}\right)$ where $L(t)$ is the length (in centimeters) of a fish t years old. [UW]

   a. What is the length of a newborn halibut at birth?

   b. Use the formula to estimate the length of a 6–year–old halibut.

   c. At what age would you expect the halibut to be 120 cm long?

   d. What is the practical (physical) significance of the number 200 in the formula for $L(t)$?

37. A cancer cell lacks normal biological growth regulation and can divide continuously. Suppose a single mouse skin cell is cancerous and its mitotic cell cycle (the time for the cell to divide once) is 20 hours. The number of cells at time t grows according to an exponential model. [UW]
    a. Find a formula $C(t)$ for the number of cancerous skin cells after t hours.
    b. Assume a typical mouse skin cell is spherical of radius 50×10^{-4} cm. Find the combined volume of all cancerous skin cells after t hours. When will the volume of cancerous cells be 1 cm^3?

38. A ship embarked on a long voyage. At the start of the voyage, there were 500 ants in the cargo hold of the ship. One week into the voyage, there were 800 ants. Suppose the population of ants is an exponential function of time. [UW]
    a. How long did it take the population to double?
    b. How long did it take the population to triple?
    c. When were there be 10,000 ants on board?
    d. There also was an exponentially growing population of anteaters on board. At the start of the voyage there were 17 anteaters, and the population of anteaters doubled every 2.8 weeks. How long into the voyage were there 200 ants per anteater?

39. The populations of termites and spiders in a certain house are growing exponentially. The house contains 100 termites the day you move in. After 4 days, the house contains 200 termites. Three days after moving in, there are two times as many termites as spiders. Eight days after moving in, there were four times as many termites as spiders. How long (in days) does it take the population of spiders to triple? [UW]

Section 4.7 Fitting Exponential Models to Data

In the previous section, we saw number lines using logarithmic scales. It is also common to see two dimensional graphs with one or both axes using a logarithmic scale.

One common use of a logarithmic scale on the vertical axis is to graph quantities that are changing exponentially, since it helps reveal relative differences. This is commonly used in stock charts, since values historically have grown exponentially over time. Both stock charts below show the Dow Jones Industrial Average, from 1928 to 2010.

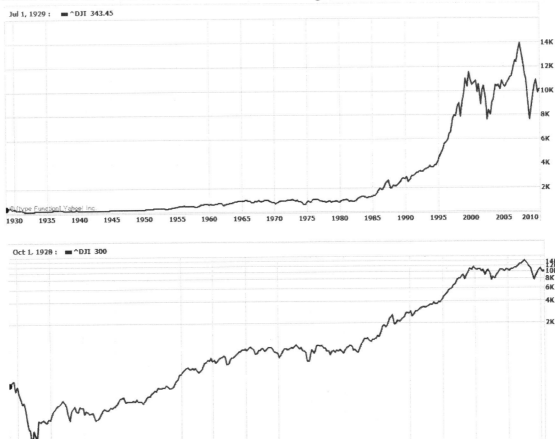

Both charts have a linear horizontal scale, but the first graph has a linear vertical scale, while the second has a logarithmic vertical scale. The first scale is the one we are more familiar with, and shows what appears to be a strong exponential trend, at least up until the year 2000.

Example 1

There were stock market drops in 1929 and 2008. Which was larger?

In the first graph, the stock market drop around 2008 looks very large, and in terms of dollar values, it was indeed a large drop. However, the second graph shows relative changes, and the drop in 2009 seems less major on this graph, and in fact the drop starting in 1929 was, percentage-wise, much more significant.

Specifically, in 2008, the Dow value dropped from about 14,000 to 8,000, a drop of 6,000. This is obviously a large value drop, and amounts to about a 43% drop. In 1929, the Dow value dropped from a high of around 380 to a low of 42 by July of 1932. While value-wise this drop of 338 is much smaller than the 2008 drop, it corresponds to a 89% drop, a much larger relative drop than in 2008. The logarithmic scale shows these relative changes.

The second graph above, in which one axis uses a linear scale and the other axis uses a logarithmic scale, is an example of a **semi-log** graph.

Semi-log and Log-log Graphs

A **semi-log** graph is a graph with one axis using a linear scale and one axis using a logarithmic scale.

A **log-log** graph is a graph with both axes using logarithmic scales.

Example 2

Plot 5 points on the graph of $f(x) = 3(2)^x$ on a semi-log graph with a logarithmic scale on the vertical axis.

To do this, we need to find 5 points on the graph, then calculate the logarithm of the output value. Arbitrarily choosing 5 input values,

x	$f(x)$	$\log(f(x))$
-3	$3(2)^{-3} = \dfrac{3}{8}$	-0.426
-1	$3(2)^{-1} = \dfrac{3}{2}$	0.176
0	$3(2)^0 = 3$	0.477
2	$3(2)^2 = 12$	1.079
5	$3(2)^5 = 96$	1.982

Plotting these values on a semi-log graph,

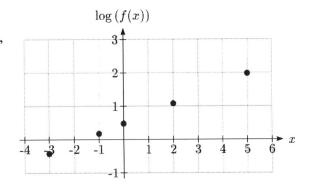

Notice that on this semi-log scale, values from the exponential function appear linear. We can show this behavior is expected by utilizing logarithmic properties. For the function $f(x) = ab^x$, finding log($f(x)$) gives

$\log(f(x)) = \log(ab^x)$ Utilizing the sum property of logs,

$\log(f(x)) = \log(a) + \log(b^x)$ Now utilizing the exponent property,

$\log(f(x)) = \log(a) + x\log(b)$

This relationship is linear, with log(a) as the vertical intercept, and log(b) as the slope. This relationship can also be utilized in reverse.

Example 3

An exponential graph is plotted on semi-log axes. Find a formula for the exponential function $g(x)$ that generated this graph.

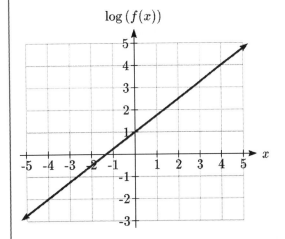

The graph is linear, with vertical intercept at (0, 1). Looking at the change between the points (0, 1) and (4, 4), we can determine the slope of the line is $\frac{3}{4}$. Since the output is log($g(x)$), this leads to the equation $\log(g(x)) = 1 + \frac{3}{4}x$.

We can solve this formula for $g(x)$ by rewriting in exponential form and simplifying:

$\log(g(x)) = 1 + \dfrac{3}{4}x$ Rewriting as an exponential,

$g(x) = 10^{1 + \frac{3}{4}x}$ Breaking this apart using exponent rules,

$g(x) = 10^{1} \cdot 10^{\frac{3}{4}x}$ Using exponent rules to group the second factor,

$g(x) = 10^{1} \cdot \left(10^{\frac{3}{4}}\right)^{x}$ Evaluating the powers of 10,

$g(x) = 10(5.623)^{x}$

Try it Now

1. An exponential graph is plotted on a semi-log graph below. Find a formula for the exponential function $g(x)$ that generated this graph.

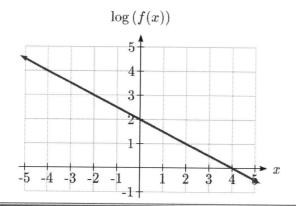

$\log(f(x))$

Fitting Exponential Functions to Data

Some technology options provide dedicated functions for finding exponential functions that fit data, but many only provide functions for fitting linear functions to data. The semi-log scale provides us with a method to fit an exponential function to data by building upon the techniques we have for fitting linear functions to data.

To fit an exponential function to a set of data using linearization
1. Find the log of the data output values
2. Find the linear equation that fits the (input, log(output)) pairs. This equation will be of the form $\log(f(x)) = b + mx$.
3. Solve this equation for the exponential function $f(x)$

Example 4

The table below shows the cost in dollars per megabyte of storage space on computer hard drives from 1980 to 2004, and the data is shown on a standard graph to the right, with the input changed to years after 1980.

Year	Cost per MB
1980	192.31
1984	87.86
1988	15.98
1992	4
1996	0.173
2000	0.006849
2004	0.001149

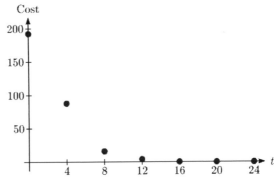

This data appears to be decreasing exponentially. To find a function that models this decay, we would start by finding the log of the costs.

Year	t	Cost per MB	log(Cost)
1980	0	192.31	2.284002
1984	4	87.86	1.943791
1988	8	15.98	1.203577
1992	12	4	0.60206
1996	16	0.173	-0.76195
2000	20	0.006849	-2.16437
2004	24	0.001149	-2.93952

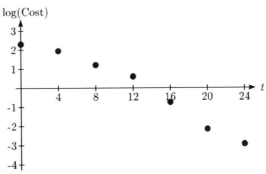

As hoped, the graph of the log of costs appears fairly linear, suggesting an exponential function will fit the original data will fit reasonably well. Using technology, we can find a linear equation to fit the log(Cost) values. Using t as years after 1980, linear regression gives the equation:
$$\log(C(t)) = 2.794 - 0.231t$$

Solving for $C(t)$,
$$C(t) = 10^{2.794 - 0.231t}$$
$$C(t) = 10^{2.794} \cdot 10^{-0.231t}$$
$$C(t) = 10^{2.794} \cdot \left(10^{-0.231}\right)^t$$
$$C(t) = 622 \cdot (0.5877)^t$$

This equation suggests that the cost per megabyte for storage on computer hard drives is decreasing by about 41% each year.

Using this function, we could predict the cost of storage in the future. Predicting the cost in the year 2020 ($t = 40$):

$C(40) = 622(0.5877)^{40} \approx 0.000000364$ dollars per megabyte, a really small number. That is equivalent to $0.36 per terabyte of hard drive storage.

Comparing the values predicted by this model to the actual data, we see the model matches the original data in order of magnitude, but the specific values appear quite different. This is, unfortunately, the best exponential model that can fit the data. It is possible that a non-exponential model would fit the data better, or there could just be wide enough variability in the data that no relatively simple model would fit the data any better.

Year	Actual Cost per MB	Cost predicted by model
1980	192.31	622.3
1984	87.86	74.3
1988	15.98	8.9
1992	4	1.1
1996	0.173	0.13
2000	0.006849	0.015
2004	0.001149	0.0018

Try it Now

2. The table below shows the value V, in billions of dollars, of US imports from China t years after 2000.

year	2000	2001	2002	2003	2004	2005
t	0	1	2	3	4	5
V	100	102.3	125.2	152.4	196.7	243.5

This data appears to be growing exponentially. Linearize this data and build a model to predict how many billions of dollars of imports were expected in 2011.

Important Topics of this Section

Semi-log graph

Log-log graph

Linearizing exponential functions

Fitting an exponential equation to data

Try it Now Answers

1. $g(x) = 10^{2-0.5x} = 10^2 \left(10^{-0.5}\right)^x$. $f(x) = 100(0.3162)^x$

2. $V(t) = 90.545(1.2078)^t$. Predicting in 2011, $V(11) = 722.45$ billion dollars

Section 4.7 Exercises

Graph each function on a semi-log scale, then find a formula for the linearized function in the form $\log(f(x)) = mx + b$.

1. $f(x) = 4(1.3)^x$ 2. $f(x) = 2(1.5)^x$

3. $f(x) = 10(0.2)^x$ 4. $f(x) = 30(0.7)^x$

The graph below is on a semi-log scale, as indicated. Find a formula for the exponential function $y(x)$.

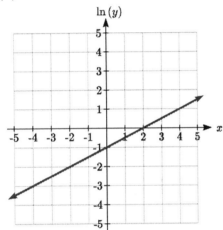

5.

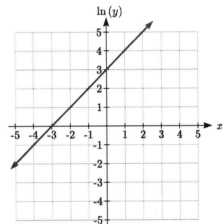

6.

7.

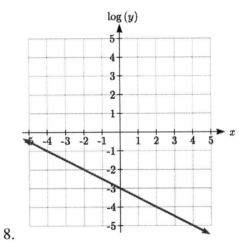

8.

Use regression to find an exponential function that best fits the data given.

9.

x	1	2	3	4	5	6
y	1125	1495	2310	3294	4650	6361

10.

x	1	2	3	4	5	6
y	643	829	920	1073	1330	1631

11.

x	1	2	3	4	5	6
y	555	383	307	210	158	122

12.

x	1	2	3	4	5	6
y	699	701	695	668	683	712

13. Total expenditures (in billions of dollars) in the US for nursing home care are shown below. Use regression to find an exponential function that models the data. What does the model predict expenditures will be in 2015?

Year	1990	1995	2000	2003	2005	2008
Expenditure	53	74	95	110	121	138

14. Light intensity as it passes through water decreases exponentially with depth. The data below shows the light intensity (in lumens) at various depths. Use regression to find an function that models the data. What does the model predict the intensity will be at 25 feet?

Depth (ft)	3	6	9	12	15	18
Lumen	11.5	8.6	6.7	5.2	3.8	2.9

15. The average price of electricity (in cents per kilowatt hour) from 1990 through 2008 is given below. Determine if a linear or exponential model better fits the data, and use the better model to predict the price of electricity in 2014.

Year	1990	1992	1994	1996	1998	2000	2002	2004	2006	2008
Cost	7.83	8.21	8.38	8.36	8.26	8.24	8.44	8.95	10.40	11.26

16. The average cost of a loaf of white bread from 1986 through 2008 is given below. Determine if a linear or exponential model better fits the data, and use the better model to predict the price of a loaf of bread in 2016.

Year	1986	1988	1990	1995	1997	2000	2002	2004	2006	2008
Cost	0.57	0.66	0.70	0.84	0.88	0.99	1.03	0.97	1.14	1.42

Chapters 5-9 (pages 337-640) form Volume 2 of this book, and can be read online at http://www.opentextbookstore.com/precalc/, or purchased as a separate printed text.

Solutions to Selected Exercises

Chapter 1

Section 1.1

1. a. $f(40) = 13$

   b. 2 Tons of garbage per week is produced by a city with a population of 5,000.

3. a. In 1995 there are 30 ducks in the lake

   b. In 2000 there are 40 ducks in the late

5. a ,b, d, e   7. a, b   9. a, b, d

11. b   13. b, c, e, f   15. $f(1) = 1$, $f(3) = 1$

17. $g(2) = 4$, $g(-3) = 2$ 19. $f(3) = 53$, $f(2) = 1$

	$f(-2)$	$f(-1)$	$f(0)$	$f(1)$	$f(2)$
21.	8	6	4	2	0
23.	49	18	3	4	21
25.	4	-1	0	1	-4
27.	4	4.414	4.732	5	5.236
29.	-4	-6	-6	-4	0
31.	5	DNE	-3	-1	-1/3
33.	1/4	1/2	1	2	4

35. a. -6   b.-16   37. a. 5   b. $-\dfrac{5}{3}$

39. a. iii   b. viii c. I   d. ii   e. vi   f. iv   g. v   h. vii

41. a. iv   b. ii   c. v   d. I   e. vi   f. iii

43. $(x-3)^2 + (y+9)^2 = 36$

45. (a)   (b)   (c)

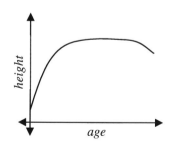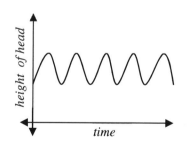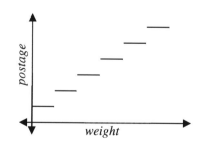

47a. t b. a c. r d. L: (c, t) and K: (a, p)

642

Section 1.2

1. D: $[-5, 3)$ R: $[0,2]$ 3. D: $2 < t \le 8$ R: $6 \le g(t) < 8$

5. D: $[0,4]$ R: $[-3, 0]$ 7. $[2,\infty)$ 9. $(-\infty,3]$

11. $(-\infty,6)\cup(6,\infty)$ 13. $\left(-\infty,-\dfrac{1}{2}\right)\cup\left(-\dfrac{1}{2},\infty\right)$

15. $[-4,4)\cup(4,\infty)$ 17. $(-\infty,-11)\cup(-11,2)\cup(2,\infty)$

	$f(-1)$	$f(0)$	$f(2)$	$f(4)$
19.	-4	6	20	34
21.	-1	-2	7	5
23.	-5	3	3	16

25. $f(x)=\begin{cases} 2 & if & -6 \le x \le -1 \\ -2 & if & -1 < x \le 2 \\ -4 & if & 2 < x \le 4 \end{cases}$ 27. $f(x)=\begin{cases} 3 & if & x \le 0 \\ x^2 & if & x > 0 \end{cases}$

29. $f(x)=\begin{cases} \dfrac{1}{x} & if & x < 0 \\ \sqrt{x} & if & x \ge 0 \end{cases}$

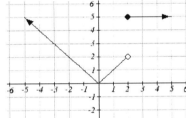

31.

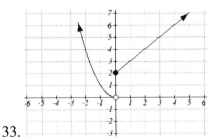

33.

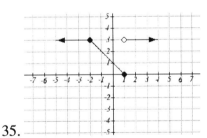

35.

Section 1.3

1. a) 6 million dollars per year   b) 2 million dollars per year

3. $\dfrac{4-5}{4-1}=-\dfrac{1}{3}$ 5. 6

7. 27

9. $\dfrac{352}{27}$

11. $4b+4$

13. 3

15. $-\dfrac{1}{13h+169}$

17. $9+9h+3h^2$

19. $4x+2h$

21. Increasing: $(-1.5, 2)$. Decreasing: $(-\infty, -1.5) \cup (2, \infty)$

23. Increasing: $(-\infty, 1) \cup (3, 4)$. Decreasing: $(1, 3) \cup (4, \infty)$

25. Increasing, concave up

27. Decreasing, concave down

29. Decreasing, concave up

31. Increasing, concave down

33. Concave up $(-\infty, 1)$. Concave down $(1, \infty)$. Inflection point at (1, 2)

35. Concave down $(-\infty, 3) \cup (3, \infty)$

37. Local minimum at (3, -22).
Inflection points at (0,5) and (2, -11).
Increasing on $(3, \infty)$. Decreasing $(-\infty, 3)$
Concave up $(-\infty, 0) \cup (2, \infty)$. Concave down $(0, 2)$

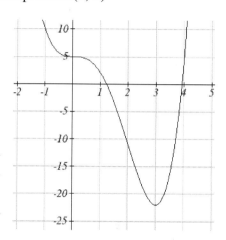

39. Local minimum at (-2, -2)
Decreasing $(-3, -2)$
Increasing $(-2, \infty)$
Concave up $(-3, \infty)$

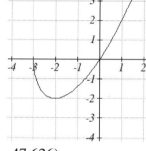

41. Local minimums at (-3.152, -47.626)
and (2.041, -32.041)
Local maximum at (-0.389, 5.979)
Inflection points at (-2, -24) and (1, -15)
Increasing $(-3.152, -0.389) \cup (2.041, \infty)$
Decreasing $(-\infty, -3.152) \cup (-0.389, 2.041)$
Concave up $(-\infty, -2) \cup (1, \infty)$
Concave down $(-2, 1)$

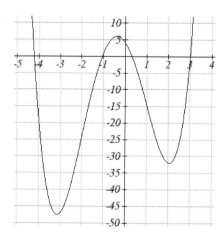

Section 1.4

1. $f(g(0)) = 36$. $g(f(0)) = -57$

3. $f(g(0)) = 4$. $g(f(0)) = 4$

5. 4   7. 9   9. 4   11. 7   13. 0   15. 4   17. 3   19. 2

21. $f(g(x)) = \dfrac{x}{7}$ $\qquad\qquad\qquad$ $g(f(x)) = 7x - 36$

23. $f(g(x)) = x + 3$ $\qquad\qquad\qquad$ $g(f(x)) = \sqrt{x^2 + 3}$

25. $f(g(x)) = |5x + 1|$ $\qquad\qquad\qquad$ $g(f(x)) = 5|x| + 1$

27. $f(g(h(x))) = (\sqrt{x} - 6)^4 + 6$

29. b $\qquad\qquad\qquad$ 31a. $r(V(t)) = \sqrt[3]{\dfrac{3(10 + 20t)}{4\pi}}$ b. 4.609in

33. $(0, \infty)$ $\qquad$ 35. $\left(-\infty, \dfrac{1}{3}\right) \cup \left(\dfrac{1}{3}, 1\right) \cup (1, \infty)$ $\qquad$ 37. $[2,5) \cup (5, \infty)$

39. $g(x) = x + 2, f(x) = x^2$ $\qquad$ 41. $f(x) = \dfrac{3}{x}, g(x) = x - 5$

43. $f(x) = 3 + \sqrt{x}, g(x) = x - 2$, or $f(x) = 3 + x, g(x) = \sqrt{x - 2}$

45a. $f(f(x)) = a(ax + b) + b = (a^2)x + (ab + b)$

b. $g(x) = \sqrt{6}x - \dfrac{8}{\sqrt{6} + 1}$ or $g(x) = -\sqrt{6}x - \dfrac{8}{1 - \sqrt{6}}$

47a. $C(f(s)) = \dfrac{70\left(\dfrac{s}{60}\right)^2}{10 + \left(\dfrac{s}{60}\right)^2}$ $\qquad$ b. $C(g(h)) = \dfrac{70(60h)^2}{10 + (60h)^2}$

c. $v(C(m)) = \dfrac{5280}{3600}\left(\dfrac{70m^2}{10 + m^2}\right)$

Section 1.5

1. Horizontal shift right 49 units
3. Horizontal shift left 3 units

5. Vertical shift up 5 units
7. Vertical shift down 2 units

9. Horizontal shift right 2 units, Vertical shift up 3 units

11. $f(x+2)+1 = \sqrt{x+2}+1$
13. $f(x-3)-4 = \dfrac{1}{x-3}-4$

15. $g(x) = f(x-1), \quad h(x) = f(x)+1$

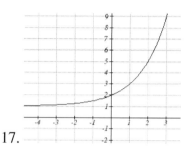
17.

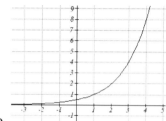
19.

21.

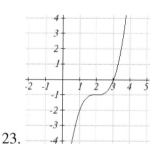
23.

25. $y = |x-3|-2$
27. $y = \sqrt{x+3}-1$
29. $y = -\sqrt{x}$

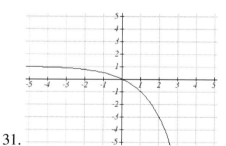
31.

33a. $-f(-x) = -6^{-x}$
    b. $-f(x+2)-3 = -6^{x+2}-3$

35. $y = -(x+1)^2 + 2$
37. $y = \sqrt{-x}+1$

39a. Even   b. Neither   c. Odd

41. Reflect *f(x)* about the *x*-axis

43. Vertically stretch *y* values by 4

45. Horizontally compress *x* values by 1/5

47. Horizontally stretch *x* values by 3

49. Reflect *f(x)* about the *y*-axis and vertically stretch *y* values by 3

51. $f(-4x) = |-4x|$

53. $\dfrac{1}{3} f(x+2) - 3 = \dfrac{1}{3(x+2)^2} - 3$

55. $f(2(x-5)) + 1 = (2(x-5))^2 + 1$

57. Horizontal shift left 1 unit, vertical stretch y values by 4, vertical shift down 5 units

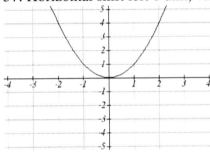 becomes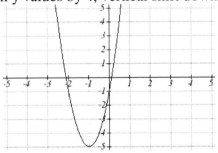

59. Horizontal shift right 4 units, vertical stretch y values by 2, reflect over x axis, vertically shift up 3 units.

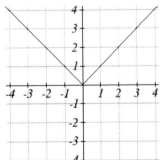 becomes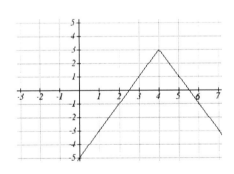

61. Vertically compress y values by ½

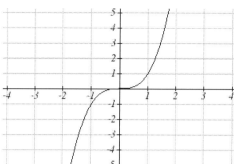 becomes

63. Horizontally stretch x values by 3, vertical shift down 3 units

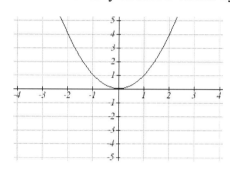 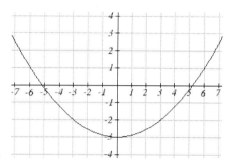 becomes

65. Reflected over the y axis, horizontally shift right 4 units $a(x) = \sqrt{-(x-4)}$

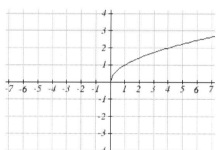 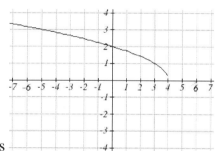 becomes

67. This function is increasing on $(-1, \infty)$ and decreasing on $(-\infty, -1)$

69. This function is decreasing on $(-\infty, 4)$

71. This function is concave down on $(-3, \infty)$ and concave up on $(-\infty, -3)$

73. This function is concave up everywhere

75. $f(-x)$

77. $3f(x)$

79. $2f(-x)$

81. $2f\left(\dfrac{1}{2}x\right)$

83. $2f(x) - 2$

85. $-f(x+1) + 3$

87. $y = -2(x+2)^2 + 3$

89. $y = \left(\dfrac{1}{2}(x-1)\right)^3 + 2$

91. $y = \sqrt{2(x+2)} + 1$

93. $y = \dfrac{-1}{(x-2)^2} + 3$ 95. $y = -2|x+1| + 3$ 97. $y = \sqrt[3]{-\dfrac{1}{2}(x-2)} + 1$

99. $f(x) = \begin{cases} (x+3)^2 + 1 & \text{if } x \le -2 \\ \dfrac{1}{2}|x-2| + 3 & \text{if } x > -2 \end{cases}$

101. $f(x) = \begin{cases} 1 & \text{if } x < -2 \\ -2(x+1)^2 + 4 & \text{if } -2 \le x \le 1 \\ \sqrt[3]{x-2} + 1 & \text{if } x > 1 \end{cases}$

103a. *Domain* : $3.5 \le x \le 6$ d. *Range* : $-9 \le y \le 7$

Section 1.6

1. 6   3. -4   5. ½

7a. 3   b. 2   c. 2   d. 3

9a. 0   b. 7   c. 1   d. 3

11.

x	1	4	7	12	16
$f^{-1}(x)$	3	6	9	13	14

13. $f^{-1}(x) = x - 3$ 15. $f^{-1}(x) = -x + 2$ 17. $f^{-1}(x) = \dfrac{x-7}{11}$

19. Restricted domain $x \ge -7, f^{-1}(x) = \sqrt{x} - 7$

21. Restricted domain $x \ge 0, \ f^{-1}(x) = \sqrt{x+5}$

23a. $f(g(x)) = \left(\sqrt[3]{x+5}\right)^3 - 5 = x$ b. $g(f(x)) = \sqrt[3]{x^3 - 5 + 5} = x$

c. This means that they are inverse functions (of each other)

Chapter 2

Section 2.1

1. $P(t) = 1700t + 45000$ 3. $D(t) = 10 + 2t$ 5. $M(n) = 40 - 2n$

7. Increasing   9. Decreasing   11. Decreasing

13. Increasing   15. Decreasing   17. 3

19. $-\dfrac{1}{3}$ 21. $\dfrac{4}{5}$ 23. $\dfrac{2}{3}$

25. -0.05 mph   (or 0.05 miles per hour toward her home)

27. Population is decreasing by 400 people per year

29. Monthly charge in dollars has an initial base charge of \$24, and increases by \$0.10 for each minute talked

31. Terry started at an elevation of 3,000 ft and is descending by 70ft per second.

33. $y = \dfrac{3}{5}x - 1$ 35. $y = 3x - 2$ 37. $y = -\dfrac{1}{3}x + \dfrac{11}{3}$

39. $y = -1.5x - 3$ 41. $y = \dfrac{2}{3}x + 1$ 43. $y = -2x + 3$

45. $P(n) = -0.004n + 34$

47. The 1st, 3rd & 4th tables are linear: respectively

　　　1. $g(x) = -3x + 5$ 3. $f(x) = 5x - 5$ 4. $k(x) = 3x - 2$

49a. $C = \dfrac{5}{9}F - \dfrac{160}{9}$ b. $F = \dfrac{9}{5}C + 32$ c. $-9.4°F$

Section 2.2

1. E   3. D   5. B

7.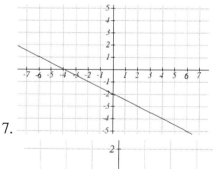

9.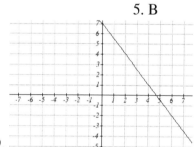

11.

13.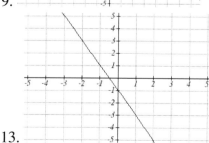

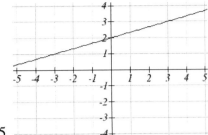

15.

17.

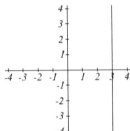

19.

21.

23. a. $g(x) = \frac{3}{4}(x+2) - 4$ b. ¾   c. -5/2

25. $y = 3$

27. $x = -3$

	Vertical Intercept	Horizontal Intercept
29.	(0,2)	(2,0)
31.	(0,-5)	(5/3, 0)
33.	(0,4)	(-10,0)

35. Line 1: $m = -10$ Line 2: $m = -10$ Parallel

37. Line 1: $m = -2$ Line 2: $m = 1$ Neither

39. Line 1: $m = -\frac{2}{3}$ Line 2: $m = \frac{3}{2}$ Perpendicular

41. $y = -5x - 2$ 43. $y = \frac{1}{2}t + 1$ 45. (-1,1)

47. (1.2, 10)   49. Plan B saves money if the miles are $> 111\frac{1}{9}$

51. $f(x) = \begin{cases} 2x+3 & if & -3 \le x < -1 \\ x-1 & if & -1 \le x \le 2 \\ -2 & if & 2 < x \le 5 \end{cases}$

Section 2.3

1a. 696 people   b. 4 years   c. 174 people per year

d. 305 people   e. $P(t) = 305 + 174t$ f. 2219 people.

3a. $C(x) = 0.15x + 10$

b. The flat monthly fee is $10 and there is an additional $0.15 fee for each additional minute used

c. $113.05

5a. $P(t) = 190t + 4170$ b. 6640 moose

7a. $R(t) = 16 - 2.1t$ b. 5.5 billion cubic feet   c. During the year 2017

9. More than 133 minutes   11. More than $42,857.14 worth of jewelry

13. 20.012 square units   15. 6 square units

17. $A = -\dfrac{b^2}{2m}$

19a. Hawaii   b. $80,640   c. During the year 1933

21. 26.225 miles

Section 2.4

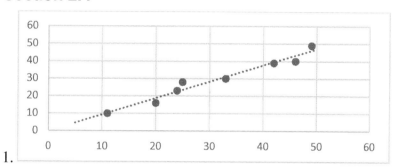

1.

3. $y = 1.971x - 3.519,\ r = 0.967$ 5. $y = -0.901x + 26.04,\ r = -0.968$

7. $17.483 \approx 17\ situps$ 9. D   11. A

13. Yes, trend appears linear because $r = 0.994$ and will exceed 35% near the end of the year 2019.

Section 2.5

1. $y = \dfrac{1}{2}|x + 2| + 1$ 3. $y = -3|x - 3| + 3$

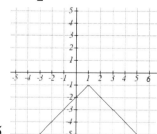

5.

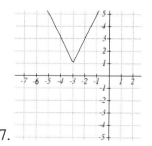

7.

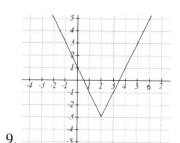

9.

11. $x = -\dfrac{9}{5}$ or $x = \dfrac{13}{5}$ 13. $x = \dfrac{1}{2}$ or $x = \dfrac{15}{2}$

15. $x = -\dfrac{5}{3}$ or $x = -\dfrac{1}{3}$

	Horizontal Intercepts	Vertical Intercept
17.	(-6, 0) and (4, 0)	(0, -8)
19.	none	(0, -7)

21. $-11 < x < 1$ or $(-11, 1)$

23. $x \geq 5$, $x \leq -1$ or $(-\infty, -1] \cup [5, \infty)$

25. $-\dfrac{13}{3} < x < -\dfrac{5}{3}$ or $(-\dfrac{13}{3}, -\dfrac{5}{3})$

Chapter 3

Section 3.1

1. As $x \to \infty$, $f(x) \to \infty$ As $x \to -\infty$, $f(x) \to \infty$

3. As $x \to \infty$, $f(x) \to \infty$ As $x \to -\infty$, $f(x) \to -\infty$

5. As $x \to \infty$, $f(x) \to -\infty$ As $x \to -\infty$, $f(x) \to -\infty$

7. As $x \to \infty$, $f(x) \to -\infty$ As $x \to -\infty$, $f(x) \to \infty$

9. 7th Degree, Leading coefficient 4

11. 2nd Degree, Leading coefficient -1

13. 4th Degree, Leading coefficient -2

15. 3rd Degree, Leading coefficient 6

17. As $x \to \infty$, $f(x) \to -\infty$ As $x \to -\infty$, $f(x) \to -\infty$

19. As $x \to \infty$, $f(x) \to \infty$ As $x \to -\infty$, $f(x) \to \infty$

21. intercepts: 5, turning points: 4   23. 3

25. 5   27. 3   29. 5

31. Horizontal Intercepts (1,0), (-2, 0), (3, 0)   Vertical Intercept (0, 12)

33. Horizontal Intercepts (1/3, 0) (-1/2, 0)   Vertical Intercept (0, 2)

Section 3.2

1. $f(x) = (x-2)^2 - 3$ 3. $f(x) = -2(x-2)^2 + 7$ 5. $f(x) = \dfrac{1}{2}(x-3)^2 - 1$

	Vertex	Vertical Intercept	Horizontal Intercepts	
7.	$(-2.5, -0.5)$	(0,12)	(-2, 0)	(-3, 0)
9.	$(2.5, -8.5)$	(0,4)	(0.438, 0)	(4.562,0)
11.	$(0.75, 1.25)$	(0,-1)	(0.191, 0)	(1.309, 0)

13. $f(x) = (x-6)^2 - 4$ 15. $f(x) = 2(x+2)^2 - 18$ 17. $b = 32$ and $c = -39$

19. $f(x) = -\dfrac{2}{3}(x+3)(x-1)$ 21. $f(x) = \dfrac{3}{5}(x-2)(x-5)$

23. $f(x) = -\dfrac{1}{4}(x-4)^2$ 25. $f(x) = -\dfrac{1}{9}(x+3)^2 + 2$

27a. 234m   b. 2909.561 ft   c. 47.735 seconds

29a. 3 ft   b. 111 ft   c. 72.497 ft

31. 24.91 in by 24.91 in

33. $125\,ft$ by $83\dfrac{1}{3}\,ft$

35. 24.6344 cm

37. $10.70

Section 3.3

C(t)	C, intercepts	t, intercepts
1.	(0,48)	(4,0), (-1,0), (6,0)
3.	(0,0)	(0,0), (2,0), (-1,0)
5.	(0,0)	(0,0), (1,0), (3,0)

7. (-1.646, 0) (3.646, 0) (5,0)

9. As $t \to \infty$, $h(t) \to \infty$ $t \to -\infty$, $h(t) \to -\infty$

11. As $t \to \infty$, $p(t) \to -\infty$ $t \to -\infty$, $p(t) \to -\infty$

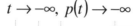

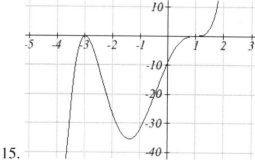

13.

15.

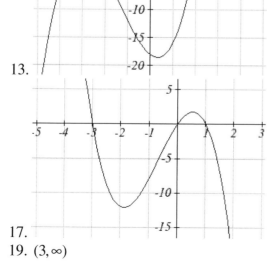

17.

19. $(3, \infty)$ 21. $(-\infty, -2) \cup (1,3)$

23. $[3.5, 6]$

25. $(-\infty, 1] \cup [4, \infty)$

27. $[-2, -2] \cup [3, \infty)$

29. $(-\infty, -4) \cup (-4, 2) \cup (2, \infty)$

31. $y = -\dfrac{2}{3}(x+2)(x-1)(x-3)$

33. $y = \dfrac{1}{3}(x-1)^2(x-3)^2(x+3)$

35. $y = -15(x-1)^2(x-3)^3$

37. $y = \dfrac{1}{2}(x+2)(x-1)(x-3)$

39. $y = -(x+1)^2(x-2)$

41. $y = -\dfrac{1}{24}(x+3)(x+2)(x-2)(x-4)$

43. $y = \dfrac{1}{24}(x+4)(x+2)(x-3)^2$

45. $y = \dfrac{1}{12}(x+2)^2(x-3)^2$

47. $y = \dfrac{1}{6}(x+3)(x+2)(x-1)^3$

49. $y = -\dfrac{1}{16}(x+3)(x+1)(x-2)^2(x-4)$

51. Base 2.58, Height 3.336

Section 3.4

1. $4x^2 + 3x - 1 = (x-3)(4x+15) + 44$

3. $5x^4 - 3x^3 + 2x^2 - 1 = (x^2+4)(5x^2 - 3x - 18) + (12x + 71)$

5. $9x^3 + 5 = (2x-3)\left(\dfrac{9}{2}x^2 + \dfrac{27}{4}x + \dfrac{81}{8}\right) + \dfrac{283}{8}$

7. $(3x^2 - 2x + 1) = (x-1)(3x+1) + 2$

9. $(3 - 4x - 2x^2) = (x+1)(-2x-2) + 5$

11. $(x^3 + 8) = (x+2)(x^2 - 2x + 4) + 0$

13. $(18x^2 - 15x - 25) = \left(x - \dfrac{5}{3}\right)(18x + 15) + 0$

15. $(2x^3 + x^2 + 2x + 1) = \left(x + \dfrac{1}{2}\right)(2x^2 + 2) + 0$

17. $(2x^3 - 3x + 1) = \left(x - \dfrac{1}{2}\right)\left(2x^2 + x - \dfrac{5}{2}\right) - \dfrac{1}{4}$

19. $(x^4 - 6x^2 + 9) = (x - \sqrt{3})(x^3 + \sqrt{3}x^2 - 3x - 3\sqrt{3}) + 0$

21. $x^3 - 6x^2 + 11x - 6 = (x-1)(x-2)(x-3)$

23. $3x^3 + 4x^2 - x - 2 = 3\left(x - \dfrac{2}{3}\right)(x+1)^2$

25. $x^3 + 2x^2 - 3x - 6 = (x+2)(x+\sqrt{3})(x-\sqrt{3})$

27. $4x^4 - 28x^3 + 61x^2 - 42x + 9 = 4\left(x - \dfrac{1}{2}\right)^2(x-3)^2$

Section 3.5

1. All of the real zeros lie in the interval $[-7,7]$
- Possible rational zeros are ± 1, ± 2, ± 3

3. All of the real zeros lie in the interval $[-13,13]$
- Possible rational zeros are ± 1, ± 2, ± 3, ± 4, ± 6, ± 12

5. All of the real zeros lie in the interval $[-8,8]$
- Possible rational zeros are ± 1, ± 7

7. All of the real zeros lie in the interval $[-3,3]$
- Possible rational zeros are $\pm \dfrac{1}{17}$, $\pm \dfrac{2}{17}$, $\pm \dfrac{5}{17}$, $\pm \dfrac{10}{17}$, ± 1, ± 2, ± 5, ± 10

9. All of the real zeros lie in the interval $\left[-\dfrac{14}{3}, \dfrac{14}{3}\right]$
- Possible rational zeros are $\pm \dfrac{1}{3}$, $\pm \dfrac{2}{3}$, $\pm \dfrac{5}{3}$, $\pm \dfrac{10}{3}$, ± 1, ± 2, ± 5, ± 10

11. $x = -2$, $x = 1$, $x = 3$ (each has mult. 1)
13. $x = -2$ (mult. 2), $x = 1$ (mult. 1), $x = 3$ (mult. 1)
15. $x = 7$ (mult. 1)
17. $x = \dfrac{5}{17}$, $x = \pm\sqrt{2}$ (each has mult. 1)
19. $x = -2$, $x = \dfrac{3 \pm \sqrt{69}}{6}$ (each has mult. 1)
21. $x = 0$, $x = \dfrac{5 \pm \sqrt{61}}{18}$ (each has mult. 1)
23. $x = \pm\sqrt{3}$ (each has mult. 1)
25. $x = \pm\sqrt{5}$ (each has mult. 1)
27. $x = \sqrt[3]{-2} = -\sqrt[3]{2}$, $x = \sqrt[3]{5}$ (each has mult. 1)
29. $x = 2$, $x = \pm\sqrt{2}$ (each has mult. 1)
31. $x = -4$ (mult. 3), $x = 6$ (mult. 2)

Section 3.6

1. $3i$

3. -12

5. $1 + \sqrt{3}i$

7. $8 - i$

9. $-11 + 4i$

11. $-12 + 8i$

13. $30 - 10i$

15. $11 + 10i$

17. 20

19. $\dfrac{3}{2}+2i$

21. $\dfrac{3}{2}+\dfrac{5}{2}i$

23. $-\dfrac{1}{25}-\dfrac{18}{25}i$

25. $f(x)=x^2-4x+13=(x-(2+3i))(x-(2-3i))$.   Zeros: $x=2\pm 3i$

27. $f(x)=3x^2+2x+10=3\left(x-\left(-\dfrac{1}{3}+\dfrac{\sqrt{29}}{3}i\right)\right)\left(x-\left(-\dfrac{1}{3}-\dfrac{\sqrt{29}}{3}i\right)\right)$.   Zeros: $x=-\dfrac{1}{3}\pm\dfrac{\sqrt{29}}{3}i$

29. $f(x)=x^3+6x^2+6x+5=(x+5)(x^2+x+1)=(x+5)\left(x-\left(-\dfrac{1}{2}+\dfrac{\sqrt{3}}{2}i\right)\right)\left(x-\left(-\dfrac{1}{2}-\dfrac{\sqrt{3}}{2}i\right)\right)$

   Zeros: $x=-5,\ x=-\dfrac{1}{2}\pm\dfrac{\sqrt{3}}{2}i$

31. $f(x)=x^3+3x^2+4x+12=(x+3)\left(x^2+4\right)=(x+3)(x+2i)(x-2i)$.   Zeros: $x=-3,\ \pm 2i$

33. $f(x)=x^3+7x^2+9x-2=(x+2)\left(x-\left(-\dfrac{5}{2}+\dfrac{\sqrt{29}}{2}\right)\right)\left(x-\left(-\dfrac{5}{2}-\dfrac{\sqrt{29}}{2}\right)\right)$

   Zeros: $x=-2,\ x=-\dfrac{5}{2}\pm\dfrac{\sqrt{29}}{2}$

35. $f(x)=4x^4-4x^3+13x^2-12x+3=\left(x-\dfrac{1}{2}\right)^2\left(4x^2+12\right)=4\left(x-\dfrac{1}{2}\right)^2(x+i\sqrt{3})(x-i\sqrt{3})$

   Zeros: $x=\dfrac{1}{2},\ x=\pm\sqrt{3}i$

37. $f(x)=x^4+x^3+7x^2+9x-18=(x+2)(x-1)\left(x^2+9\right)=(x+2)(x-1)(x+3i)(x-3i)$

   Zeros: $x=-2,\ 1,\ \pm 3i$

39.

$f(x)=-3x^4-8x^3-12x^2-12x-5=(x+1)^2\left(-3x^2-2x-5\right)\ =-3(x+1)^2\left(x-\left(-\dfrac{1}{3}+\dfrac{\sqrt{14}}{3}i\right)\right)\left(x-\left(-\dfrac{1}{3}-\dfrac{\sqrt{14}}{3}i\right)\right)$

   Zeros: $x=-1,\ x=-\dfrac{1}{3}\pm\dfrac{\sqrt{14}}{3}i$

41. $f(x)=x^4+9x^2+20=\left(x^2+4\right)\left(x^2+5\right)=(x-2i)(x+2i)\left(x-i\sqrt{5}\right)\left(x+i\sqrt{5}\right)$

   Zeros: $x=\pm 2i,\ \pm i\sqrt{5}$

Section 3.7

1. D   3. A

	Vertical Asymptotes	Horizontal Asymptote	Vertical y-Intercept	Horizontal x-intercept

5.	$x = -4$	$y = 2$	$(0,-3/4)$	$(3/2, 0)$
7.	$x = 2$	$y = 0$	$(0,1)$	DNE
9.	$x = -4, 1\frac{1}{3}$	$y = 1$	$(0, 5/16)$	$(-1/3, 0), \ (5,0)$
11.	$x = -1$, hole at $x = 1$	$y = 1$	$(0,3)$	$(-3, 0)$
13.	$x = 4$	none $y=2x$ (oblique)	$(0, ¼)$	$(-1, 0), (1/2, 0)$
15.	$x = 0, \ 4$	$y = 0$	DNE	$(-2, 0), (2/3, 0)$
17.	$x = -2, \ 4$	$y = 1$	$(0, -15/16)$	$(1, 0), (-3, 0), (5, 0)$

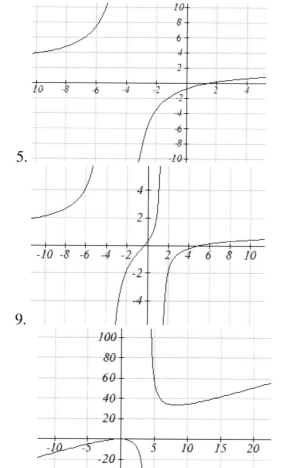

5.

9.

13.

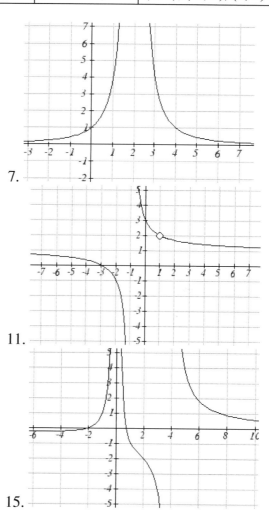

7.

11.

15.

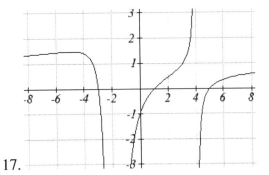

17.

19. $y = \dfrac{50(x-2)(x+1)}{(x+5)(x-5)}$

21. $y = \dfrac{7(x-4)(x+6)}{(x+4)(x+5)}$

23. $y = \dfrac{1(x-2)^2}{2(x+1)}$

25. $y = \dfrac{4(x-3)}{(x+3)(x-4)}$

27. $y = \dfrac{27(x-2)}{(x+3)(x-3)^2}$

29. $y = \dfrac{1(x+3)(x-2)}{3(x-1)}$

31. $y = \dfrac{-6(x-1)^2}{(x+3)(x-2)^2}$

33. $y = -\dfrac{2(x)(x-3)}{(x+3)(x-4)}$

35. $y = \dfrac{2(x-1)^3}{(x+1)(x-2)^2}$

37. $y = \dfrac{(x-4)(x-2)}{(x-4)(x+1)}$

39. $y = 3x - 2$

41. $y = \dfrac{1}{2}x + 1$

43. $y = -2x + 1$

45. a. $C(n) = \dfrac{4}{20+n}$ b. $C(10) \approx 13.33\%$ c. 80 mL d. as $n \to \infty, C \to 0$

Section 3.8

1. Domain $(4, \infty)$ Inverse $f^{-1}(x) = \sqrt{x} + 4$

3. Domain $(-\infty, 0)$ Inverse $f^{-1}(x) = -\sqrt{12 - x}$

5. Domain $(-\infty, \infty)$ Inverse $f^{-1}(x) = \sqrt[3]{\dfrac{x-1}{3}}$

7. $f^{-1}(x) = \dfrac{(x-9)^2}{4} + 1$

9. $f^{-1}(x) = \left(\dfrac{x-9}{2}\right)^3$

11. $f^{-1}(x) = \dfrac{2-8x}{x}$

13. $f^{-1}(x) = \dfrac{3-7x}{x-1}$

15. $f^{-1}(x) = \dfrac{5x-4}{3+4x}$

17. 65.574 mph

19. 34.073 mph

21. 14.142 feet

Chapter 4

Section 4.1

1. Linear   3. Exponential   5. Neither

7. $P(t) = 11,000(1.085)^t$ 9. 47622 Fox

11. \$17561.70   13. $y = 6(5)^x$ 15. $y = 2000(0.1)^x$

17. $y = 3(2)^x$ 19. $y = \left(\frac{1}{6}\right)^{-\frac{3}{5}}\left(\frac{1}{6}\right)^{\frac{x}{5}} = 2.93(0.699)^x$ 21. $y = \frac{1}{8}(2)^x$

23. $34.32\,mg$ 25. 1.39%; \$155,368.09   27. \$4,813.55

29. Annual $\approx$ \$7353.84   Quarterly $\approx$ \$7,469.63   Monthly $\approx$ \$7,496.71

Continuously $\approx$ \$7,510.44

31. 3.03% 33. 7.4 years

35a. $w(t) = (1.113)(1.046)^t$ b. \$1.11   c. Below what the model predicts $\approx$ \$5.70

Section 4.2

1. B   3. A   5. E   7. D   9. C

11. 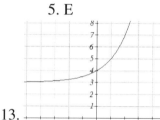   13.   15.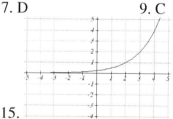

17. $y = 4^x + 4$ 19. $y = 4^{x+2}$ 21. $y = -4^x$

23. As $x \to \infty$ $f(x) \to -\infty$. As $x \to -\infty$ $f(x) \to -1$

25. As $x \to \infty$ $f(x) \to -2$ As $x \to -\infty$ $f(x) \to \infty$

27. As $x \to \infty$ $f(x) \to 2$ As $x \to -\infty$ $f(x) \to \infty$

29. $y = -2^{x+2} + 1 = -4(2)^x + 1$ 31. $y = -2(2)^{-x} + 3$

33. $y = -2(3)^x + 7$ 35. $y = 2\left(\frac{1}{2}\right)^x - 4$

Section 4.3

1. $4^m = q$ 3. $a^c = b$ 5. $10^t = v$

7. $e^n = w$ 9. $\log_4(y) = x$ 11. $\log_c(k) = d$

13. $\log(b) = a$ 15. $\ln(h) = k$ 17. 9

19. 1/8   21. 1000   23. e^2

25. 2   27. -3   29. ½

31. 4

33. -3

35. -2

37. -1.398

39. 2.708

41. $\dfrac{\log(14)}{\log(5)} \approx 1.6397$

43. $\dfrac{\log\left(\frac{1}{15}\right)}{\log(7)} \approx -1.392$

45. $\dfrac{\ln(17)}{5} \approx 0.567$

47. $\dfrac{\frac{\log(38)}{\log(3)}+5}{4} \approx 2.078$

49. $\dfrac{\log(5)}{\log(1.03)} \approx 54.449$

51. $\dfrac{\log\left(\frac{8}{3}\right)}{3\log(1.04)} \approx 8.335$

53. $\dfrac{\ln\left(\frac{1}{5}\right)}{-0.12} \approx 13.412$

55. $\dfrac{\log\left(\frac{5}{8}\right)}{\log\left(\frac{1}{2}\right)} \approx 0.678$

57. $f(t) = 300e^{-0.0943t}$

59. $f(t) = 10e^{0.03922t}$

61. $f(t) = 150(1.0618)^t$

63. $f(t) = 50(0.98807)^t$

65. During the year 2013

67. During the year 2074

69. $\approx 34\,hours$

71. 13.532 years

Section 4.4

1. $\log_3(4)$

3. $\log_3(7)$

5. $\log_3(5)$

7. $\log_7(2)$

9. $\log(6x^9)$

11. $\ln(2x^7)$

13. $\log\left(x^2(x+1)^3\right)$

15. $\log\left(\dfrac{xz^3}{\sqrt{y}}\right)$

17. $15\log(x)+13\log(y)-19\log(z)$

19. $-2\ln(a)+4\ln(b)-5\ln(c)$

21. $\dfrac{3}{2}\log(x)-2\log(y)$

23. $\ln(y)+\dfrac{1}{2}\left(\ln(y)-\ln(1-y)\right)$

25. $\dfrac{8}{3}\log(x)+\dfrac{14}{3}\log(y)$

27. $x \approx -0.717$

29. $x \approx -6.395$

31. $t \approx 17.329$

33. $x = \dfrac{2}{7}$

35. $x \approx 0.123$

37. $x \approx 4.642$

39. $x \approx 30.158$

41. $x \approx -2.889$.

43. $x \approx 6.873\,\text{or}\; x \approx -0.873$

45. $x = \dfrac{12}{11} \approx 1.091$

47. $x = 10$

Section 4.5

1. Domain: $: x > 5$ V. A. @ $x = 5$

3. Domain: $x < 3$ V.A. @ $x = 3$

5. Domain: $x > -\dfrac{1}{3}$ V.A. @ $x = -\dfrac{1}{3}$

7. Domain: $x < 0$ V.A. @ $x = 0$

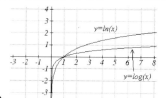

9.

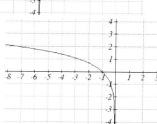

11.

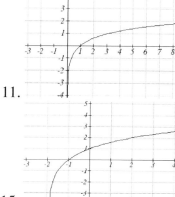

13.

15.

17. $y = \dfrac{1}{\log(2)}\log(-(x-1))$

19. $y = -\dfrac{3}{\log(3)}\log(x+4)$

21. $y = \dfrac{3}{\log(4)}\log(x+2)$

23. $y = -\dfrac{2}{\log(5)}\log(-(x-5))$

Section 4.6

1. $f(t) = 13(0.9195)^t$.   2 mg will remain after 22.3098 minutes

3. $f(t) = 200(0.999564)^t$. $f(1000) = 129.3311$ mg

5. $r = -0.06448$.   Initial mass: 9.9018 mg.   After 3 days: 0.01648 mg

7. $f(t) = 250(0.9909)^t$.   Half-life = 75.8653 minutes

9. $f(t) = a(0.999879)^t$.   60% $(0.60a)$ would remain after 4222.813 years

11. $P(t) = 1500(1.02337)^t$ (t in minutes).   After 2 hours = 24000.

    After 100 minutes = 15119

13. a) 610.5143 (about 611)   b) 25.6427 minutes   c) 10431.21   d) 106.9642 minutes

15. 23.1914 years

17. 53.319 hours

19. $T(t) = 90(0.99166)^t + 75$.   a) 134.212 deg   b) 112.743 minutes

662

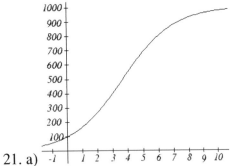

21. a) 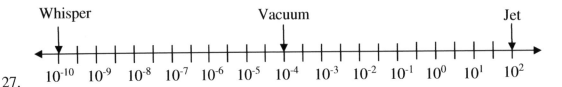 b) 100 c) 269.487 d) 7.324 years

23. $\log(x) = -0.5$. $x = 0.3162$ 25. $\log(x) = 1.5$. $x = 31.623$

27.

| Whisper | | | | | | Vacuum | | | | | | Jet |

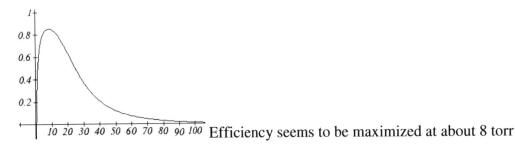

10^{-10} 10^{-9} 10^{-8} 10^{-7} 10^{-6} 10^{-5} 10^{-4} 10^{-3} 10^{-2} 10^{-1} 10^{0} 10^{1} 10^{2}

29. 63095.7 times more intense   31. MMS magnitude 5.817

33. a) about 1640671 b) 1.4 hours c) No, because $(2.042727)^{0.693147} \approx e^{0.495105}$

     d) Anja's data predicts a continuous growth rate of 0.4116, which is much smaller

     than the rate 0.495105 you calculated. Our model would overestimate the number of

     cells.

35. a) The curve that increases rapidly at first is $M(p)$

     b) $H(100) = 0.9775$

     c) Myoglobin: $M(20) = 0.9524$. Hemoglobin: $H(20) = 0.3242$

     d) At 20 torrs: 0.6282. At 40 torrs: 0.2060. At 60 torrs: 0.0714

     Efficiency seems to be maximized at about 8 torr

37. a) $C(t) = 1.03526^{t}$, or $C(t) = e^{0.03466t}$

     b) Volume of one cell: $\frac{4}{3}\pi(50\times10^{-4})^{3} \approx 5.236\times10^{-7}$ cm^3, so will need about

     1.9099×10^{6} cells for a volume of 1cm^3. $C(t) = 1.9099\times10^{6}$ after 417.3 hours

39. 31.699 days

Section 4.7

1. $\log(f(x)) = \log(1.3)x + \log(4)$

3. $\log(f(x)) = \log(0.2)x + 1$

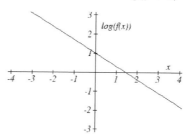

5. $y = e^{\frac{1}{2}x-1} = e^{-1}e^{\frac{1}{2}x} \approx 0.368(1.6487)^x$

7. $y = 10^{-x-2} = 10^{-2}10^{-1x} = 0.01(0.1)^x$

9. $y = 776.682(1.426)^x$ 11. $y = 731.92(0.738)^x$

13. Expenditures are approximately \$205

15. $y = 7.599(1.016)^x$ $r = 0.83064$, $y = 0.1493x + 7.4893$, $r = 0.81713$. Using the better function, we predict electricity will be 11.157 cents per kwh

Answers for Chapter 5-9 exercises can be found in the second volume of this text.

Index